Charles Pitfield Mitchell

Dissolution and Evolution and the Science of Medicine

An attempt to co-ordinate the necessary facts of pathology and to establish the

first principles of treatment

Charles Pitfield Mitchell

Dissolution and Evolution and the Science of Medicine
An attempt to co-ordinate the necessary facts of pathology and to establish the first principles of treatment

ISBN/EAN: 9783337312466

Printed in Europe, USA, Canada, Australia, Japan

Cover: Foto ©berggeist007 / pixelio.de

More available books at **www.hansebooks.com**

DISSOLUTION AND EVOLUTION

AND THE

SCIENCE OF MEDICINE.

DISSOLUTION AND EVOLUTION

AND THE

SCIENCE OF MEDICINE:

AN ATTEMPT TO CO-ORDINATE THE NECESSARY FACTS OF
PATHOLOGY AND TO ESTABLISH THE FIRST
PRINCIPLES OF TREATMENT.

Sanitarium

BY

C. PITFIELD MITCHELL,

MEMBER OF THE ROYAL COLLEGE OF SURGEONS, ENGLAND;
AUTHOR OF 'THE TREATMENT OF WOUNDS AS BASED ON EVOLUTIONARY LAWS.'

LONDON:

LONGMANS, GREEN, AND CO.

AND NEW YORK: 15 EAST 16ᵗʰ STREET.

1888.

PREFACE.

By the following pages it is proposed to disseminate some new applications of Mr. Herbert Spencer's leading generalisations. The sustaining elements of the Synthetic Philosophy are the doctrines of evolution and dissolution. The design is to inquire whether these may not be made fertilising principles for large collections of the data of pathology, and thus the means of practice for the physician and surgeon.

The raw material of medical science grows with an acceleration of rate that gives acuteness to the need of great central truths about which facts may be organised. I venture to think that the doctrines of dissolution and evolution supply, in useful measure, this large and pressing want ; and, contrary to all appearances, that they bear, in their present application, upon every aspect of the work of the practitioner. In substantiation thereof this volume is offered, but here the prospect must be unfolded.

A multitude of facts will be seen to acquire a

remarkable concinnity, unity, and order. The unification of truths so voluminous and diverse as those
constituting pathology is undoubtedly of educational
value. To make all diseases, from a whitlow to
mania, one in principle by cause and effect is an
aid to practical thought. It vastly augments the
carrying capacity of the mind. I am convinced that
one of the chief hindrances to our advancement to
an organised system of rational procedure in the
treatment of disease is the want of a perception of
the uniformities that exist amid the diversities of
morbid phenomena. The proclivity, handed down
to us from pre-scientific times, to regard each disease
as individual, is discernible almost everywhere in the
scientific medicine of to-day. With primitive man,
diseases were personified entities; with ourselves, they
are pathological entities. It is surely as desirable
for the practical physician as for the philosopher to
perceive the continuity and inseparableness of natural
processes and their most essential general differences
and likenesses. General truths that will break down
the factitious divisions which convention has raised
are exceedingly helpful instruments in the progress
of science and art. I believe that the principles of
dissolution and evolution are such general truths.

But it is among the services of great generalisations to exert a purifying influence upon all that
comes within the range of their generalising power.
If true, they are at variance with what is untrue, and

will lead to the correction of erroneous observation
and erroneous inference. An influence of this kind
may be seen in the co-ordination of the data of
pathology by means of Mr. Spencer's resplendent
discoveries. To mention first some of the minor
effects of this influence. There is the idea, suggested
by the formula of dissolution, of inflammation as a
process of disintegration throughout the series of its
essential component phenomena. In truthfulness,
comprehensiveness, and simplicity, the conception
approaches perfection, and immediately clarifies all
previous notions of the pathology of inflammation.

There are the so-called interstitial inflammations known as ' fibroses,' ' cirrhoses,' and 'scleroses.'
Dissolution and evolution at once hint that the increase of connective tissue in these morbid changes
is the result of a process the opposite of inflammatory, the result of a reparative process—the connective-tissue growth is a scar. When the changes are
further examined by the light of this suggestion, and
by further light from the general principles, it is
found that every consideration consists with the
view. That it will ultimately become the prevailing
view I have not the least doubt, and its acceptance
at the present juncture would modify momentously
our ways of looking at the chief organic pathological
processes.

I will not particularise the new aspects in which
the principles present the phenomena of regressive

metamorphosis, of tumour-formation, of the organic diseases, of functional diseases, and of insanity; but will observe that each of these aspects should assist in shaping the actions of the practitioner. Only among leading physicians is it appreciated that a knowledge of the intimate changes of diseases, and our manner of viewing them, are more than practice. They are its very mainsprings, and, still more, have the governance or direction of it; not only do they supply the motive power, but they make practice good or bad. Let us suppose that the case for which advice is sought is one of croup, where laryngeal obstruction threatens to induce death. One skilled in practice but imperfectly acquainted with the recondite and interdependent chemical and physical tissue-changes which are the essence of the disease, will advise and perform tracheotomy, or intubation of the larynx; or pin himself to medication, poultices, and steam inhalations. But such a one will be heedless of the remote causes of the asphyxia. The urine may be examined and albumen discovered, but beyond adding to the gravity of the case, the albuminuria will not have much significance; it will not serve as a clue to the antecedents of the laryngeal obstruction. These are not known, they will not be thought of, and no steps will be taken to deal with them either by removing the conditions upon which they depend, or by furthering the natural processes of functional adjustment. The case will probably terminate fatally

by extension of the disease to the bronchi, by septicæmia, or nephritis. Hence, in a medical journal of
to-day we read that of fifty cases of croup treated
by the new method, 'intubation of the larynx,' death
resulted in thirty-eight. In the cases of the twelve
patients that survived, it is not certain that intubation was the only factor concerned in the re-
coveries.[1]

There are results of pre-eminent importance issu-
ing from the application of dissolution and evolution
to pathology. I refer to the proposition that diseases
are the outcome of interactions between the organism
and its environment. This is not an unfamiliar proposition. Ziegler says: 'A true disease is not a consequence of the indwelling and inherited properties of
the cell. The efficient causes of a disease are always
external. In the observations we made on the amœba,
it was heat or cold, an altered surrounding medium,
or the galvanic current, which brought about disease
and death. All these noxious influences are derived
from without; and what we have here remarked in a
single instance experience shows us to be universal.
Autonomous as the cell may seem, it is yet unable,
without external impulsion, to heighten its functions
above the physiological standard, or, on the other
hand, to check or to suppress them. We can, therefore, give a still more exact definition of the notion

---

[1] I shall not be understood to inveigh against tracheotomy, intubation of the larynx, or inhalations as measures of expediency.

of disease. By the term disease we are to understand
a deviation of some of the vital manifestations from
the normal, the deviation being conditioned by external
influences.'[1] Now this conception, perspicuous enough
when it applies to the vital manifestations of the
simplest organisms, loses all clearness and is without
effect upon practical procedure when we come to the
diseases of man. But, seeing that the induction is
all-important both in the prevention and cure of
disease, it needs to be established and brought into
the boldest relief by instance upon instance endlessly
varied. It is needed to *show*, if possible, that every
disease is the outcome of environmental actions, and
that every symptom also expresses a relation of the
organism to external conditions. As dissolution is
here applied to pathology, it is proved, nearly as far
as it is at present possible to prove, that the deepest
causes of all diseases lie outside the organism. To
us, the most sterling merit of the formula of dissolution is that it holds the mind in constant communion
with the external factors in pathogenesis.

To the question, Are diseases inherited ? as important as any in practical medicine, the doctrines of
dissolution and evolution will be found to reflect an
answer as unexpected as it is, I think, satisfactory.
In the final chapter it is shown to be a necessary
conclusion from many considerations that non-congenital diseases are rarely subject to heredity in the

---

[1] *A Text-book of Pathological Anatomy and Pathogenesis.*

way currently believed. Out of this conclusion—
potent for incalculable good if true—and out of a
conception I have presented of the processes of organic equilibration (organic equilibration is the *vis
medicatrix naturæ*) there emerge fundamental canons
of rational medical practice, and a medical practice
dynamic in present possibilities and of dazzling
promise.

Thus the remoteness of the subject-matter of this
work from the daily concerns of the practitioner is
only apparent.

It may possibly be thought that the attempt to
make a synthesis of all orders of the facts of pathology is rather supererogatory ; that a less exhaustive
treatment would perhaps have been sufficient. The
monotony of a single theme long drawn out is certainly very fatiguing ; but I think the partial or
general exemplification of principles embracing so
wide a diversity of fact and idea would not avail to
obtain an extensive recognition for the principles.
Unlimited reiteration is necessary that their truth and
power may penetrate many minds ; and is necessary
also for the acquirement of an easy use of them
among new orders of phenomena. Who, for example,
ignorant of histological morbid anatomy, but knowing something of gross anatomy, would assent to the
mere statement that the consolidated lung of pneumonia is a partially disintegrated lung ; that cancer

results from a process of disintegration in the tissues ; or that tuberculosis of the hip-joint is related to the action of external forces ? Yet these statements are true beyond all doubt. It is with clear design that I have carried out the illustrations in plenteous detail.

I am conscious that the argument is in very many places most incomplete and defectively presented. The phraseology is often far from being what I wish, and it has caused me much regret that many interesting points have had to be dealt with in a very superficial manner. But the extent of the ground to be covered, and my own deficiencies, have rendered it difficult to produce a more satisfactory work.

I would that the task I have undertaken had fallen into hands more competent than my own to do justice to Mr. Spencer and others whose work I have availed myself of. As must be obvious to anyone familiar with Mr. Spencer's writings, my essay is to the Synthetic Philosophy as an absolutely dependent outgrowth ; and if it has other than commonplace merits, the credit in large measure properly belongs to Mr. Spencer. The reader may, however, feel assured that upon the work of Mr. Spencer, as well as upon the work of others, I have exercised such truth-testing capacity as I possess.

The task has occupied me while embarked in private practice. I have therefore been afforded opportunity to apply the supreme proof of medical doctrine, opinion, and statement : that of their congruity or incongruity with personal experience in the observation and treatment of disease. It is needful to say that the interpretations herein given of the Synthetic Philosophy are my own, and that the responsibility attached thereto rests solely upon me.

With comparatively few exceptions, only the material of recently-published text-books has been dealt with, the statements of fact resting mainly on the authority of their general acceptance by the profession. Consequently the text is free from the incubus of many references.

For the sake of completeness, I have included a few illustrations from nervous and mental diseases ; but with unwillingness to trespass upon the field occupied by Dr. Hughlings Jackson, M. Ribot, Dr. Henry Maudsley, Dr. James Ross, and Dr. Charles Mercier.

While residing latterly in New York I have been so fortunate as to enlist the able offices of Dr. Hermann M. Biggs, of the Carnegie Laboratory. He has assisted me in satisfying myself upon many points of morbid anatomy, and has supplied me with much valuable suggestion. For these services my thanks are heartily tendered to him.

The following are the text-books from which I have drawn most freely : Dr. T. H. Green's ' Introduction to Pathology and Morbid Anatomy'; Dr. Joseph Coats's 'Manual of Pathology'; Ziegler's 'Text-book of Pathological Anatomy and Pathogenesis,' translated by Dr. Donald MacAlister; Dr. Klein's 'Elements of Histology'; and 'Insanity and Allied Neuroses,' by Dr. George H. Savage. I have found Dr. Quain's 'Dictionary of Medicine' inexpressibly serviceable. The 'Text-book of the Principles of Physics,' by Mr. Alfred Daniell, M.A., and the 'Outlines of Psychology,' by Mr. James Sully, M.A., have been very helpful. I am much beholden to Mr. Bland Sutton, F.R.C.S., for the chapters on Teratomata and Cystomata in his recent work, 'An Introduction to General Pathology.' Figures 1 and 5 I owe to Dr. Coats's 'Pathology,' and figures 2, 3, and 4, to the text-book of Dr. Green.

If the views offered upon the nature and origin of tumours have the value I suppose, then it should be said that their development has been greatly furthered by the writings of Mr. J. F. Payne, M.B., F.R.C.P., and the utterances of the distinguished participants in the memorable 'Discussion on Cancer.'

These, and the acknowledgments in the body of the work, are a very inadequate requital of my obligations to contemporary writers. It is possible to give only a fragmentary history of the mental savings of several years.

LONDON : *March* 1888.

CONTENTS.

DISSOLUTION AND EVOLUTION

AND THE

SCIENCE OF MEDICINE.

——•◇•——

INTRODUCTORY.

HERBERT SPENCER has brought many to the conviction that his view of the processes of nature as divi-
sible into evolutional and dissolutional processes is
essentially veracious. If the formulas which he has
given to us, setting forth the salient features of these
processes, are placed in apposition with any group of
natural changes coming within the purview of observation, there is discovered an invariable correspondence between the formulas and the facts they purport
to represent. It does appear from examination that
all the changes going on within and without us have
the characters of the one or the other order of mutations, and that the generalisations, in their most
abstract form, really stand for universal verities.
Therefore it may be possible to translate the facts of
any science in terms of these all-embracing principles. In our endeavour to do this with the data of
pathology the mode of exposition will be to take the

h   B formulated statements of the principles and compare with them the facts they are expected to unify. First, the essential truths of general pathology, as those of inflammation, suppuration, repair, and resolution, will be considered in respect of their conformability to the formulas. Next, the retrograde metamorphoses and neoplasms will be looked at from the same points of view, later chapters including a rough inspection of the general pathological changes as they are presented in special diseases. Finally, the results will be considered from practical aspects.

This manner of presentation seems to possess superiority in simplicity and directness, and in permitting a comprehensive survey of the field. The difficulties are of some magnitude in the way of making intelligible application of doctrines as elaborate in organisation as the doctrines of evolution and dissolution ; and any other method of presentation would, perhaps, have involved an undesirable discursiveness of exposition.

Some acquaintance with 'First Principles' will be required for the easy understanding of all that follows ; but I am in hope that the simple arrangement adopted will make the work acceptable to medical readers unfamiliar with the Synthetic Philosophy.

PART I.

GENERAL DISEASE.

CHAPTER I.

§ 1. PRELIMINARY.

Definition I.—'Dissolution is a disintegration of matter and concomitant absorption of motion; during which the matter passes from a definite, coherent heterogeneity to an indefinite, incoherent homogeneity; and during which the retained motion undergoes a parallel transformation.'

The process of dissolution is antithetical to the process of evolution; and the reader will find the inductive basis of this formula expounded *per contra* in abundant detail in 'First Principles,' chapters xiii., xvii.; cf. also chapters xii. and xxiii.

As in the investigation of the phenomena called physical, changes in the living organism are made known to us as alterations in the form of the matter of an aggregate, and as alterations in the form of its energy. In the language peculiar to the science of biology these changes are spoken of as morphological and physiological. Inasmuch, however, that matter may be a species of energy, and its distinction as substance forced by the nature of our experiences, the dual division of phenomena into material and non-material may possibly be only temporary. A

higher plane of contemplation would combine the two
as one form of experience. But scientific knowledge
and conceptions have not yet reached a stage of
development permitting this; hence the cognisance
taken in the formula of changes of matter and
changes of energy, or motion, to use the term in the
formula.

For, since the varieties of energy, at least the
varieties of kinetic energy, are conceivably modes of
motion, it appears allowable to give co-extension and
connature to kinetic energy and motion ; therefore,
these terms will generally be used interchangeably.
' Scientific scrutiny, so far as it has penetrated, finds
motion throughout external nature—motions everywhere, motions underlying every phenomenon, however different from motions some of them may seem
to common apprehension ; and no *scientific* investigation has as yet detected anything but motion.' [1]

The changes of matter in the organism are chiefly
and most definitely perceived as histological transformations ; the changes of motion as molar and
functional transformations. Molar action is the visible movement of bodies, as of particles and cells—
cells undergo transpositions, and their protoplasm
is capable of intrinsic movement. What we distinguish as functional action is the *equivalent of
motion*, since, whatever the form of the function, we
may think of it as proceeding from the motion of
molecules, whether such function is manifested as
perceptible motion, heat, chemical change, or the
operations of the mind and nervous system.

[1] J. Johnstone Stoney, F.R.S., in *Nature*, xxxi. 529.

§ 2. The Disintegration of Matter.

We shall not expect the law of dissolution, formulated as above, to obtain striking exemplification in the phenomena presented by inflammation of liminal intensity. Inflammation being, as we now know, merely a perversion of normal functions and structures, the transition from normality to abnormality must necessarily be gradational. And this however swiftly the transition is effected. Thus, there is a stage or degree of inflammation which closely approximates to health, and one which widely deviates from it; and where the deviation is extreme we may look for the plainest marks of dissolution. Yet even in incipient inflammation faint though certain marks can, I think, be descried.

At the outset the reader may be reminded that the formulas do not stand for state or being, but for changes of state, and are therefore to be understood always in relative senses, that is, in connection with antecedent being. The transformation in evolution, for example, is not to absolute but to relative definiteness, coherence, and heterogeneity. Hence, in seeing whether the anatomical and physiological changes characteristic of inflammation are comprehended by the principle of dissolution, reference must be made to the state we call normal, and it must be conceived from our present point of view.

What is the normal arrangement of matter and motion in a compound structure such as inflammation in the human subject commonly involves? The

cellular and non-cellular units of every tissue-system
have an *integrated* arrangement due to the interrelations of their component forces. In simple vascular
connective tissue, which may be conveniently selected
for consideration, the blood-vessels, lymph-vessels,
nerves, and connective tissue cells are all bound
together, and are interdependent both structurally
and functionally. Not only are the stationary elements united in close aggregation, but the moving
elements are constrained to take definite directions.
The fluids have their courses prescribed by the vessels, and the blood-corpuscles traverse distinct paths.
For the most part, as when the circulation is active,
the corpuscles are disposed centrally in blood-vessels
while the blood-plasma flows at the periphery.

The blood-cells and plasma having in certain
directions definitely limited space relations like the
fixed components of the tissue, if these space relations
are so disturbed that corpuscles and plasma overstep
their normal confines, they become less integrated, or
relatively disintegrated.

We may say that a structure is being disintegrated
when, from a common centre, the motions of its units
are increasing; and is being integrated when its units
are approaching a common centre.

Inflammation is first evidenced, when its phenomena are observed by means of the microscope, by
changes complying with this definition; there is a
dispersion of corpuscles and plasma. These constituents of simple connective tissue come to occupy
more space; they begin to lose their aggregated
form, moving away from one another.

In the adjoined figure (Fig. 1) are seen on the left hand the normal configuration of blood compacted by the vessels, and on the right, over the area of inflammation, corpuscles scattered in all directions. In health the leucocytes, for the greater part assembled with the coloured discs as the core of the

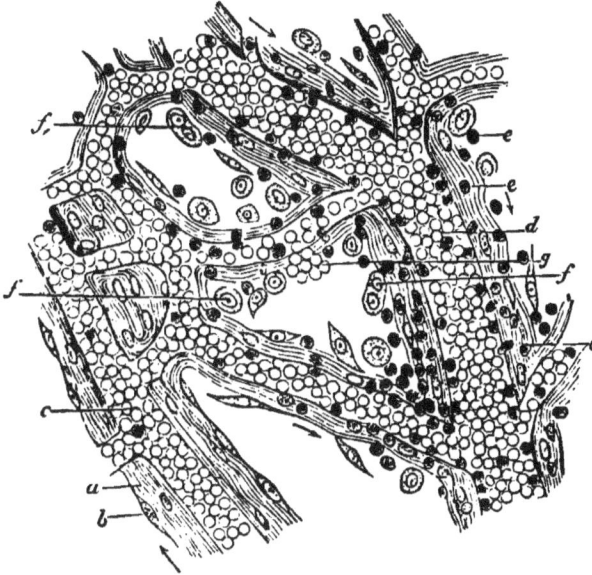

Fig. 1.—Inflamed human omentum. The phenomena of inflammation are seen in the veins and capillaries, the condition being normal at the artery (c) where b represents endothelium covering the trabecula (a). In the vein (d) there are many white corpuscles along the wall. Some of these are emigrating (e); f, desquamated endothelium; g, extravasated red corpuscles. (ZIEGLER).

vascular column, at the setting in of inflammation pass to the circumference and issue through the walls of the capillaries and veins. And there is the extravasation of red cells known as diapedesis. Now the truest and simplest conception of such disparting of units is that the collective mass of blood and vessels is undergoing disintegration.

Arterial dilatation is a primary effect of inflammatory action. Whatever the particular physiological explanation of this phenomenon, it is the beginning of disunion among the structural units of the arteries. The endothelial plates of the capillaries lose their closeness of cohesion, so that stomata are produced, and the trabecular endothelia desquamate—further disaggregations.

A more obvious disintegration is the exudation of plasma. The plasma, or certain constituents of it, is freed from its normal structural associations and escapes, going to form the well-known inflammatory products serum and fibrine. Cellular proliferation undoubtedly forms a part of degeneration and other morbid processes; and if we think it part of the process of inflammation, then it must be counted among the disintegrations. It is essentially a protoplasmic dissolution (see chap. v.). In the genesis of cells, either the parent whole is reduced to a number of parts, which is complete disruption, or particles of protoplasm are detached, as nuclei or gemmæ, and this is partial disruption. Further, as the new cells are born they do not, while the inflammation is in progress, approach or integrate towards common centres, but are diffused, wandering from place to place.

This is inflammation of moderate degree. But if we pass from incipient to advanced forms of the process there may be discriminated increasing degrees of disintegration. Through the wall of a burrowing abscess, proceeding from the healthy to the destroyed tissues, there are found the following changes. In the zone of least perceptible change there is an exudation of serum and a few leucocytes ; then the exudate becomes more coagulable—more of the blood is separated from the vessels, and red cells in large number are extruded ; next, thrombosis of vessels and stasis, with consequent softening of vessels and tissue cells, may be observed ; and at the centre complete molar disintegration, or necrosis.

Thus, the most conspicuous changes, seen microscopically, in inflammation of a simple tissue have in common this character, namely, that constituent parts are dislocated or loosened from their centres of attachment or attraction and become less integrated, or relatively disintegrated. It is reasonable, therefore, to see in the phenomena of inflammation a correspondence with the first proposition of the principle of dissolution.

Inflammation is said to be sometimes initiated by a *contraction* of vessels. As this is most probably an effect of reflex action, and not essential to the process of inflammation, it hardly makes the foregoing correspondence less complete. Nor does the massing or concentration of cells in inflammatory stasis. Here the elements of the blood may cohere simply on account of the mechanical conditions, and the absolute or relative subsidence of circulation-forces. The injury done to the vascular endothelium is said to increase the adhesiveness of the leucocytes.

The inflammatory changes just noticed are usually transitory. Hereafter we shall have to examine in connection with disintegration the permanent alterations of structure which are secondary to inflamma-

tion, and classed among the retrograde metamorphoses. Where inflammation terminates in suppuration, sloughing, gangrene, ulceration, hæmorrhage,
and the discharge of exudations, the disintegration
is very obvious. Integral parts are then destroyed,
and may be separated completely from the body.

§ 3. THE ABSORPTION OF MOTION.

By the absorption of motion in dissolution we are
to understand the imbibition of any of the varieties
of kinetic energy—thermal, chemical, mechanical, or
other (§ 1). And this absorption of energy is not
only a concomitant of the disintegration of matter,
but is also the *cause* of the disintegration. The induction is, that particles of matter come to occupy
more space from receiving additional motion from the
environment.

In the present and succeeding illustrations the
term of the formula of dissolution we are now considering should recognise, in the most general way,
the cause or causes of pathological changes.

First, let us look at the common examples of
dissolution or disintegration as associated with the
reception of an excess of energy from without.

Where inorganic bodies are broken up into their
constituent elements by distinguishable causes, the
changes are known to be wrought by the action of
surrounding forces. When heat is the force at work,
it may, either slowly and imperceptibly, or quickly
and perceptibly, reduce solid bodies to their elementary parts.

Where the molecules of gases, liquids, and solids are diffused through or dissolved by different media, the energy required for diffusion is obtained from external sources. Water evaporates speedily or tardily, other conditions being the same, in proportion to the amount of heat-energy imparted to it ; and the rapid melting of sugar, say in a warm infusion of tea, represents work done both by liquid and thermal motion in overcoming the cohesions of the molecules of the immersed solid.

' Masses of sediment accumulated into strata, afterwards compressed by many thousands of feet of superincumbent strata, and reduced in course of time to a solid state, may remain for millions of years unchanged ; but in subsequent millions of years they are inevitably exposed to disintegrating actions,' such as the actions of the waves, rain, frost, and snow.

Organic matter retains or loses its figure, in numberless familiar cases, according as it receives much or little kinetic energy in the shape of liquid, thermal, and other environmental motion. As all know, drying or freezing organic substances is preservative of them; and heat, and moisture, and micro-organisms are destructive.

How directly related to the incidence of some external force may be the disintegration of living aggregates, the following illustration shows, and it shows also the nature of the relation.

If the hand is placed for a few moments in water having a temperature of 100° F., we notice among the effects an increase in the limb's bulk or volume ;

whereas if the hand is placed in water of 40° F. its volume is lessened. Regarding these effects from a physical standpoint, and disregarding the physiological mechanisms by which they are brought about, and by which in certain cases they may be neutralised, we see in them a conformity to the rule that bodies expand when heated and contract when cooled. In explanation we may apply the theory valid in the case of inorganic bodies, that the augmented molecular activities of the tissues are such as to increase the intermolecular spaces, and therefore the total space occupancy of the limb. Now let the hand sustain contact with water of sufficiently high temperature to cause inflammation and subsequent death of tissue. In this event, the enlargement of the amplitude of molecular oscillations is such, in the first instance, as to induce that scattering of cells and fluids which, we have seen, is characteristic of inflammation ; and in the suppurative stage, the amplitude of molecular oscillation is such as to have for its resultant a molar segregation. The destroyed parts are actually dislocated, and finally dissevered and cast off from the body—either *en masse*, as slough, or in particulate form, as pus.

Regarding this dissociation of a portion of the body from our special point of view, it is discerned that the impacts of the thermal vibrations have been adequate to destroy not only the qualities distinguished as vital, but what is essential to this vitality —the union subsisting between the parts and the corporate organism. The physical fact of disunion may be perceived in its nakedness, by paralleling the case of heat applied to an inorganic body until its component particles burst asunder.

Mr. Spencer's arguments tend to show that the redistributions of energy in any system are the manifestations of forces everywhere in opposition as attractions and repulsions. That is to say, evolution or dissolution is the differential result of a play of energies, resolvable into the ultimate forms in which force is made known to us—attraction and repulsion. Therefore we may perhaps say of the foregoing case that the forces by which parts were held together have been more than counterbalanced by the repulsive tendencies of the thermal movements of the molecules.

In conclusion, it should be noted that the taking up of energy is not peculiar to dissolving aggregates. All bodies, whether integrating or disintegrating, receive molar or molecular motion from surrounding things and give off motion to them. This reciprocal transference of energy in the case of heat finds its expression in the theory of exchanges. Living creatures during the evolutional stages of their existence absorb stores of potential force in the form of food, and of actual force in the form of light and heat. The truth really enunciated by the principle of dissolution is not simply that dissolving aggregates absorb motion, that this absorption is a necessary concomitant of the disintegration of matter, but that the motion absorbed is the *cause* of the dissolution.

For, as just mentioned, evolution or dissolution is the resultant of contrary actions ; now it testifies to the ascendency of conditions favourable to a concentration of parts and a dissipation of kinetic energy, and now to the ascendency of reverse conditions—those favourable to diffusion of parts and an absorption of energy. In the words of Mr. Spencer, ' neither of these two antagonistic processes ever goes on absolutely unqualified by the other . . . a change towards either is a differential result of the conflict between them. An evolving aggregate, while on the average losing motion and integrating, is always in one way or other receiving some motion, and to that extent disintegrating.' [1] The resultant changes are not dissolutional until the energy received preponderates over the energy dissipated.

Prepared by these reflections, we may proceed to inquire whether inflammation and suppuration are attended by the conveyance of a superaddition of external energy.

When the conditions of the cases are known, inflammation may then be causatively referred to one or more of the outer influences to which the organism is exposed. If the inflammatory process is excited by mechanical injury, by chemical agents, by burning and scalding, or by parasites, the transference of energy to the organism is direct and manifest. It is less direct and less manifest where tissues are inflamed from contact with poisonous matters distributed to them by the blood ; for instance, in gout and the later lesions of syphilis. Yet here the phlogogenic agents come originally from without ; in syphilis the virus is derived from another organism, and gout is

---

[1] *First Principles*, § 177.

unquestionably dependent on dietetic and other habits, that is, upon environing actions (see chap. vii. § 3).

In many other cases the causes of particular inflammations cannot be traced even remotely to the organism's conditions. What shall we say of these? When we come to consider the involved and well-nigh inscrutable but ever interdependent pathological sequences met with in special diseases, good reasons will appear for believing that where inflammations cannot be finally followed out to the operation of external conditions the cases are imperfectly understood; and that, with great probability, were our knowledge of such cases complete, a reception of energy from without would prove to be the primary cause of all inflammations. It is impracticable to bring forth here the *a posteriori* evidence of this, so complicated and multitudinous are the facts. Therefore, the reader is referred to succeeding chapters. We may just notice that each progressive step in our knowledge of the causes of inflammation brings us nearer to the recognition of some external factor or group of factors. Mention may be made of formerly-so-called idiopathic pleurisy and peritonitis. Some of these forms of inflammation are now known to be associated in causative sequence with morbid states certainly referrible to external actions. The pleurisy secondary to Bright's disease caused by intemperance is an example, and we have another in the peritonitis secondary to metritis arising from septic infection.

Experimentally produced inflammation demonstrates unequivocally the correspondence to the formula we are seeking. When inflammation is excited by incising the web of a frog's foot, we see that the mechanical power which severs the tissues is not only the cause, but is also quantitatively related to the degree of inflammation. Other things being equal, the deeper the injury inflicted, or the larger the quantity of disturbing motion imparted to the elements of the tissue, the more intense the morbid process.

In considering (§ 2) how the matter of a part is disintegrated in inflammation, we saw that this disintegration is first indicated by a diffusedness of the organic particles. Hence, in the case now instanced, we must recognise in the centrifugal movements of cells and fluids in the web of the frog the persistence of that mechanical energy noted as the source of the disturbances. Similarly, where there is a different etiology, as when nitric acid or heat is the phlogogenic agent, the chemical or thermal energies are conserved in part, we may assume, in the visible perverted motions which result. Or perhaps a more faithful representation of the actuality is the following. In inflammation the first effect of the causative agent is to produce a partial disintegration of the vascular structure; as would be the case with an inorganic aggregate, the incident forces shake the units slightly apart. The extravasation of serum and leucocytes, and diapedesis, are the consequences of this change in the vessels.

In the causes of suppuration, the relation under consideration is daily demonstrated, and most clearly in the causes of suppuration in wounds. When caustic silver is applied to granulations, these structures will almost visibly melt into pus. Drainage-tubes, sutures, and ligatures very frequently give rise to tracts of suppurating tissue. There is also 'antiseptic suppuration,' from irritation of the antiseptic agent, sometimes witnessed in aseptic wounds treated with carbolic acid or perchloride of mercury. And putrefactive and other micro-organisms known to have external origin are among the well-established causes of suppuration; indeed, they are proclaimed by some observers to be the only cause.

Pus formation as witnessed in abscesses, either primary or metastatic, is, in many cases, known to be directly or indirectly related to some external influence. Instances are furnished by pyæmic abscesses, and by phthisical vomicæ.

Thus, as far as the facts extend, they do affirm that inflammation and suppuration are attended by an absorption of environmental energy.

It may make conception of the process of inflammation clearer if we consider here the motion disappearing when functions are annihilated. The imperceptible molecular activities of living matter, which have their summations in visible functional action—the contraction of muscles, the circulation of fluids, secretion and excretion—cease at death; and there is an analogous cessation of function in parts that are inflamed, as is well known. What becomes of this functional motion? It is not easily understood as furthering the inflammatory process. The answer appears to be contained in our conception of the attributes of organic function. What exactly should we mean by function, as we perceive it in the vital actions of simple tissue, in the contraction of muscle, or in the digestion of food ? A common quality of these and all organic functions, whether simple or compound, is that of co-ordinate action ; the constitutional energies of tissues and organs are combined for some distinct end. But in being co-ordinate, the energies bear to one another a determinate—not necessarily equal—proportion in time, space, and magnitude, *and it is this determinateness of the proportions which makes the actions what we call co-ordinate.* First, to exemplify the time ratios. The exchanges of matter and motion going on in any simple tissue are first summed up as the primitive processes of assimilation and disassimilation. These follow one another in times more or less definitely proportioned. If, following dissociation, the re-association or absorption of new material is unduly postponed, the functions of assimilation and disassimilation are without co-ordination, or come altogether to an end. Again, the blood-vessels bring food-material, and carry away the products of its decomposition ; therefore, the actions of the blood-distributing system, and associated nervous functions of this system, must be definitely timed to the processes of assimilation and disassimilation. And behind these resultant and recognisable forms of organic energy, we may infer among the factors the existence of precise ratios that are unrecognisable —ratios among the molecular forces.

That the energies of this simple tissue must act through constant spaces is implied by the uniformity of its structural conformation. In witness thereof the space relations of blood-vessels and the outlying cells to be fed by them are invariable.

The determinate character of the force-magnitudes finds illustration in the provisions for regulating blood-propulsion and cell-alimentation. If the strengths of these are not in certain proportions, if the blood-pressure is too high or too low, if aliment is unduly deficient or unduly excessive, the combined energies of the tissue will not be co-ordinate but inco-ordinate.

When we say that the function of muscle is to contract and do work we mean that its purpose (function) is subserved by contraction for some specified object. If this object is the adjustment of the calibre of vessels, the muscle-cells and fibrils must contract at certain proportionate times, through certain proportionate spaces, and with certain proportionate strengths; otherwise the object of contraction is not attained, or is attained imperfectly, inco-ordinately. And so with the simultaneous and successive actions making up the functions of digestion, or other organised group of forces : these also must be in determinate ratios.[1]

Organic function, then, being the expression of energies co-ordinate in time, space, and magnitude, an effect of any inflammation-producing agent will be to make them inco-ordinate ; the ratios among them will become indefinite, or less definite ; and the specific function will be annulled temporarily, or

[1] I am indebted to Dr. Charles Mercier for the suggestion of th views. See his essay on Inco-ordination, *Brain*, vol. vi. p. 78.

permanently. The agents causing inflammations do not, however, unless necrosis is the sequel, destroy *all* organic co-ordination, but only that which we distinguish as the special work of the cellular community. The alteration from a combination of forces in definite proportions to one in indefinite proportions is a relative one, the change being from higher co-ordinations to lower ones. While a cirrhosed kidney has almost lost that coherent arrangement of its energies which ends in the excretion of urea, there may still be going on in its substance lower grades of coherent action, as the excretion of water, the processes of cell-respiration and reproduction. Similarly with the elements of muscle that have been degraded by suppuration, or the elements of nerve in Wallerian degeneration.

If this view of the matter is acceptable, it will explain that disappearance of specialised functional motion which is an accompaniment of inflammation and death. Yet it should not be forgotten that, when any force giving origin to the inflammatory process has worked its effects, a re-doing of the part—reso·lution or repair—sets in if the conditions are favourable, and energy is probably then drafted off from neighbouring parts for the purposes of this new development. Hence there will be local deductions from the force otherwise available for normal work.

If we may not conclude from the foregoing that inflammation and suppuration are marked by a disintegration of matter and absorption of motion, a closely approaching coincidence with this clause of the formula of dissolution must be conceded

/

§ 4. The Change from a Definite (*special*), Coherent (*organised*) Heterogeneity (*multiformity*)
(*complexity*) to an Indefinite (*general*), Incoherent (*unorganised*) Homogeneity (*uniformity*)
(*simplicity*).

Besides the equivalents here introduced within
parentheses, while exhibiting the illustrations in the
present and following chapters, the terms of this
clause of the formula will receive ample definition.

When adverting just now to the inco-ordination
of function attendant upon inflammation, there was
partly anticipated the present task, which is to inquire whether functions and structures become relatively indefinite, incoherent, and homogeneous.

We may take some form of local inflammation
and suppuration—say an acute abscess, or the suppurative degeneration of granulations—where the
result is a destruction of tissue. In the *process* of
suppuration the solid elements of the tissue may
presumably contribute to the formation of the pus or
be liquefied and absorbed by pus derived from the
serum and leucocytes of the blood. But in all cases
of suppurative destruction of portions of the body
the rearrangements of matter and motion are identical
in general character.

Forcing itself first upon attention is the change
from heterogeneity to homogeneity. However multiform, morphologically and physiologically considered,
was the tissue destroyed, the resulting product—pus
—is little more than an assemblage of homogeneous

cells bathed in a pabulum of serum. And, looking further, we find that tissue-elements which were coherently associated and specialised have become less coherently associated and unspecialised both as to anatomical disposition and service or function. Let the example be a metastatic abscess of the kidney; then the glomeruli, tubules, secreting cells, vessels, and nerves, all combined into a definite structure, and doing definite work in the economy, have been deposed for simple cells having neither determinate arrangement nor functional utility, and only such combination as is afforded by a common nutrient fluid.

If the suppuration is not joined with marked destruction of the tissues, as where pus is discharged from mucous and serous surfaces, the elements of the purulent matter are degenerated blood-constituents and superficial cells of epithelium; and in this case the transformations do not answer so strikingly to the requirements of the formula. Morphologically, pus-cells bear a close likeness to white corpuscles; yet there are some structural and other dissimilarities. They do not exhibit amœboid movements, their nuclei are divided, and they are incapable of absorbing, like the leucocytes of granulations, such dead tissue as a sequestrum of bone. These attributes are understood to signify degenerative changes. Regarded from another aspect, pus leucocytes do not play in the economy the special *rôle* of blood leucocytes, and they are in no sense subordinate to or united in coherent working with the other components of tissues. So that suppuration, both when resulting in loss of solid tissue and simply in loss of elements of the blood, is in accord with dissolution, being a change to relative indefiniteness, incoherence, and homogeneity of functions and structures.

Is this clause of the formula descriptive of inflammation ending in resolution? The functions and structures of parts are decided as specialisation is advanced. In the simple tissue depicted in fig. 1, the more highly specialised organs, the blood-vessels, are seen in the inflamed portions to have lost something of their normal definiteness; there is a slight obscuration of parts in that the contrast between the vessels and the extravascular tissue is less pronounced.

Functionally considered, the blood-vessels have not their normal distinctness, inasmuch as their contents are distributed indifferently, here and there, by extravasation through the vessel-walls, and not to the normal seats of tissue-respiration and alimentation.

The diminished structural coherence and consequent inco-ordination of function scarcely require delineation. With the escape of serum and leucocytes, and with the stasis of blood, the pre-existing orderly arrangements give place to confusion.

Apparently, functional and structural heterogeneity is increased rather than lessened by the new arrangement. Greater heterogeneity, however, implies a specialisation of parts, is consequent, indeed, on specialisation; and what we really see in inflammation is an increase in the *number* of the more simple and less specialised units, the white corpuscles; and, were the inflammation to terminate destructively, these bodies, aggregated as pus, would be the only remaining representatives of the previously differentiated structures. And the previous diversity of function would give place to uniformity—simple assimilation, disassimilation, and multiplication among the leucocytes. Inflammation, then, as well as suppuration, harmonises with the final terms of the formula.

This chapter may be appropriately concluded with a short reference to local anæmia, active and passive hyperæmia, and hæmorrhage, as dissolutional changes.

Only in extreme forms, and in the structural transformations following long-continued anæmia and hyperæmia, do we find all the characters the formula denotes. As states of the vessels, both deficiency and excess of blood may occur within the boundaries of normality. The pallor of the face in fright or the turgescence of the hands after exposure to cold cannot be regarded as pathological. But, natural limits being exceeded, it is worth while to note one or two correspondences with the formula —signs of disintegration as an effect of absorbed energy.

A vessel that has been long deprived of blood, as by ligatures, loses the power to retain serum, leucocytes, and red cells ; when the circulation is restored they pass through the interstices of the vessel-walls. The previous anæmia has induced that degree of vascular disintegration we observed in inflammation. Nutrition having been stopped, the plastids (units of structure) are unable to withstand the incident motions—chemical and mechanical strains—and the slight disruption ensuing makes possible the extravasations.

The phenomena and conditions of hyperæmia are quite comparable. Relatively diminished arterial resistance, variously caused, is uniformly the morphological condition upon which arterial hyperæmia depends. This is equal to saying that the less integrated state of the vascular tissue permits the overfilling with blood and the transudation of serum sometimes associated with active hyperæmia.

Mechanical or venous congestion is always a consequence of motion conveyed in the shape of increased blood-pressure in the veins; and the resulting disintegration is borne witness to by transudation of serum, diapedesis, and hæmorrhage. Some of the enduring changes, as fibroid induration, thrombosis, and necrosis, resulting from prolonged passive hyperæmia, will be considered in later chapters.

Hæmorrhage, either as diapedesis or from the coarse rupture of vessels, is obviously a disintegration of the body. Diapedesis is due to communicated motion, as we have just seen, and in hæmorrhage from traumatic injury energy is manifestly absorbed. Of hæmorrhage associated with deteriorated blood, as in leukæmia, scurvy, or small-pox, or hæmorrhage from ulceration and atheroma, the first causes are numerous, often obscure and entangled, and the reader must be referred to the examples to be met with in the section occupied with special diseases.

CHAPTER II.

RESOLUTION AND REPAIR AS EVOLUTIONAL CHANGES.

Definition II.—' Evolution is an integration of matter and concomitant dissipation of motion ; during which the matter passes from an indefinite, incoherent homogeneity, to a definite, coherent heterogeneity ; and during which the retained motion undergoes a parallel transformation.'

§ 1. The Integration of Matter and Dissipation of Motion.

Single living particles, or their assemblages as tissues, organs, and compound individuals, present us with a conservative system of forces. Employing the terminology of physical science, if, in these assemblages, the configuration of matter and motion is deformed within certain limits the system is under stress, and the *status quo* will be restored when the deforming action ceases. If the distortion is that known as inflammation, it may be of greater or less degree, to which the subsequent restitution of figure will be related. Should the inflammation be, in the main, limited to a deflection of the courses normally taken by the fluids and moving cells, and have as its visible product simply an exudation of serum or fibrine and hæmocytes into the meshes of the tissue ; then the organic aggregate will behave in a way comparable to an inorganic aggregate strained within the limits of its elasticity ; the previous configuration will be restored with all measurable completeness. The cells and fluids will resume their habitual courses, and the exudations will become absorbed. This is resolution after inflammation.

Should the phlogistic process be of severer form, and have for its product a cellular necrosis, or other destruction of tissue, then the organic structure will behave in a way comparable to an inorganic structure strained *beyond* the limits of its elasticity ; there will have been wrought a ' permanent set.' That equilibrium may be restored, there must be a redevelopment of particles. This is repair after inflammation, involving the birth of new tissue-elements and the organisation of them into structure.

It was seen in the preceding chapter that inflammation, as observed microscopically, consists of changes corresponding to those formulated as dissolution. The correspondence was seen to be most distinct when inflammation ended in complete disintegration, sloughing, suppuration, and ulceration. Hence, resolution being the sequence of a partial inflammatory dissolution, it will, we may expect, less perspicuously agree with the formula of evolution than will repair, which is the sequence of complete dissolution. Yet that resolution and evolution are the same in principle seems sufficiently clear.

Taking, again, the inflammation of simple tissue, the fact must be borne in mind that we are dealing with a system normally in *moving* equilibrium. The units of living tissue have their own proper visible motions, and these should be kept apart from the motions which are proper to the process of resolution. We find, for example, that in inflammation blood-cells are concentrated within the vessels, and as resolution begins they gradually move off owing to the restoration of the normal blood-current. In this respect units are separated, not brought together.

It being understood that always prior to the setting in of resolution the cause of the inflammation has ceased, then in inflammation of simple tissue the fixed cells lie bathed in a leucocyte-containing exudate. Now when resolution takes place, this exudate and contained cells are taken up either by local cell-imbibition or by lymphatics and veins. There is thus a reaggregation within cells or vessels, and this constitutes integration (chap. i. § 2). With the contraction of arteries and the closing of capillary stomata there is a further consolidation of substance.

But is there in the process of renewal we are considering a concomitant dissipation of motion? Exuded serum, or plasma, and corpuscles, are matter diffused through the walls of vessels, and in becoming diffused the particles must have acquired relative motion. Diffusion is a separation of parts, and parts cannot become separated without their motions in respect of one another being increased. Therefore the motion taken up in the inflammatory diffusion of hæmocytes and plasma must be rendered again when these elements concentrate or are integrated by absorption, as in resolution. And being thus rendered again, it is dissipated or lost in the meaning of the formula, though of course, as already pointed out, motion is gained by the restoration of circulation.

So that resolution shows us a reversion of the antecedent partial dissolution, in that matter is integrated and motion dissipated.

§ 2. THE CHANGE FROM AN INDEFINITE, INCOHERENT HOMOGENEITY TO A DEFINITE, COHERENT HETEROGENEITY.

In comparing the process of resolution with the remaining terms of the definition of evolution, it is perhaps enough to say that since inflammation has been shown to be betokened by a loss of definiteness, coherence, and heterogeneity of function and structure, recovery from inflammation—resolution—must be a gain of these qualities. With the resorption of an exudate—say into pulmonary alveoli—the normal exchanges of matter and motion are re-established. Functions and structures return to the standard of speciality, co-ordination, and variety from which they had regressively diverged.

In considering whether the phenomena of repair fulfil the requirements of the formula, we may choose, as suitable in its simplicity, and as a type of reparative action, the repair of wounds by the growth of granulations.

Confining attention to the positively ascertained

phenomena of this mode of healing, there is first to be noted the deposition of leucocytes and plasma upon the wounded surface. Whether these leucocytes have origin in the fixed cells of connective tissue, or are emigrant white blood-corpuscles, need not now concern us. We are assured of the fact of their generation at the seat of an antecedent inflammation, and that by their increase in mass and structural differentiation they become connective tissue.

In the early stage of development these cells have amœboid characters, undergoing changes of shape and movements of translation which the feebleness of their mutual attachments permits. With each stage of development there is a progressive consolidation of the cells and vascular components of the young tissue, and a cessation of the changes of cell-form, and the movements from one place to another. In the final stage the resulting tissue is compact, and the non-vascular elements of it are fixed : the units have lost their visible motions.

This graduation of primitive plasma and formative cells into mature connective tissue is thus a continuous integration of matter and concomitant dissipation of motion.

That increased definiteness, coherence, and heterogeneity of matter and motion are coincident with the foregoing changes appears to be demonstrable. The development of connective tissue from granulation tissue commences in relatively homogeneous plasma and simple cells, having little organisation and undecided functions. Later the cells become spindle-shaped, their substance differentiates into fibres, blood and lymph channels are marked out as vessels, and all become firmly joined together, forming a definite and coherent structure, having definite and coherent purposes.

The phenomena of resolution and repair as witnessed in systemic diseases will obtain further notice subsequently. Here it may be observed that what has just been said respecting the reproduction of common reparative tissue—connective tissue—is true, *mutatis mutandis*, of structures of higher speciality; of the regeneration of bone, cartilage, muscle, or epithelium.

CHAPTER III.

THE RETROGRADE METAMORPHOSES AS DISSOLUTIONAL
CHANGES.

§ 1. COAGULATION OF THE BLOOD.

COAGULATION of the blood, being the essential pheno-. menon of many kinds of retrograde metamorphosis, naturally comes first in the order of the present illus- trations. But there is immediately encountered a fact which apparently negatives any expectation that co- agulation changes are generalised by the principle of dissolution (chap. i. def. 1). Instead of material dis- integration the foremost characteristic of dissolution, the death of blood is distinctly a material integration, the leading characteristic of evolution (chap. ii. def. 2). Obviously, in the formation of solid fibrine from fluid blood there is not a diffusion but a concentra- tion of parts. How is this to be reconciled with any presumption that coagulation is a dissolutional process ?

Here we are re-introduced to a truth without which the phenomena of disease will be but loosely translated by the formulas of dissolution and evo- lution. There has already been occasion to note (chap. i. § 3) that evolution or dissolution expresses the difference between forces tending to dispersion of the members of a system and forces tending to their condensation, and that in no aggregate does either of the processes go on singly. Whichever the change observed, it is a differential result. Thus, while a body in the totality of its changes may be evolving, portions of it are dissolving ; and, conversely, while the entirety of the changes may be dissolutional some of the details will be evolutional. ' Everywhere, and to the last . . . the change at any moment going on forms a part of one or the other of the two processes.' . . . ' The chances are infinity to one against these opposite changes balancing one another ; and if they do not balance one another, the aggregate as a whole is integrating or disintegrating.' [1]

Illustrations by case and example will give clear meaning to these statements. In the normal living organism the two orders of mutation are ever in process throughout the system, are mutually complementary, and jointly contributory to the total evolutional result. This is shown in assimilation and disassimilation, processes of evolution and dissolution ; in interstitial waste and repair ; in accretion and excretion. The secretions of the mammary and other glands are good examples of normal dissolutions.

In decomposition after death not only is there a reduction of organic substance to simple constituents —carbonic acid, water, and ammonia, &c.—but many chemical redintegrations are effected, as in the formation of the organic alkaloids, the ptomaines.

The chemical changes of inorganic systems supply innumerable examples. If manganese monoxide and

[1] *First Principles*, p. 285.

hydrochloric acid are mixed together under appropriate conditions, the resulting redistribution is a dissociation of chlorine gas, and an association of oxygen with hydrogen to form water, and of chlorine with manganese to form manganous chloride ($MnO_2 + 4HCl = MnCl_2 + 2H_2O + Cl_2$). Here the change in the whole is analytical, minor syntheses accompanying.

Everywhere in the domain of pathology dissolution and evolution are presented to us in this aspect. When detailing the histological changes in inflammation it was seen that, at large, the process is a dissolution of particles, but attended by incidental unions of particles. Sometimes arteries contract, and, blood stagnating, there ensue local integrations of hæmocytes. Thus with the general disaggregation there are associated particular aggregations. Throughout the present and succeeding chapters this point will demand frequent recognition. We shall meet with examples in dealing with the neoplasms; the organic diseases, as nephritis and pneumonia ; the functional diseases ; and diseases of the mind. And it is the key to understanding our present subject—coagulation as a process of dissolution. It will be seen that the integration of matter to form fibrine, in the clotting of blood, is explicable as due to a play of particular chemical attractions. According to prevailing views, these attractions are between fibrinogen dissolved in the blood-plasma and fibrino-plastin and fibrine-ferment contained in the white corpuscles. In deciding whether the phenomena of blood-coagulation satisfy the formula of dissolution by including a disaggregation of matter, the blood must be regarded as a system or group of combined elements, and then there is found to be really a physical disunion of substance. The white corpuscles and, as some assert, the blood-plaques, are in part destroyed, their destruction appearing to be a necessary condition of fibrine formation ; and the unformed element, the plasma, is broken up into simpler bodies—the liquor sanguinis and fibrinogen ; the secondary integration of fibrinogen and fibrino-plastin being quite in harmony with the general doctrine of dissolution.

But do the conditions of coagulation show its relation to an accession of environing energy ?

Blood clots slowly if the circumstances permit the reception of little motion, and, on the other hand, clotting takes place quickly if much motion is received. Attesting this, blood kept in contact with living tissue is inapt to coagulate. 'If an artery be ligatured in two places and cut out while full of blood it may be hung up and the blood will remain fluid for some days.' Clearly the blood is here unfavourably situated for receiving any form of molar or molecular energy with which it is not equipoised or that is competent to overthrow the balance of its contained forces. The blood and tissue together form an equilibrated system.

When blood is exposed to the chemical, thermal, mechanical, or other energies present, say, in the atmosphere, with which it is not organically balanced, coagulation is facile. Contact with the air, whipping the blood, the introduction of needles into vessels, or the destruction of vascular endothelium all suffice to induce clotting. In these ways there are imparted to the blood motions that cannot be harmonised with its own intrinsic motions, and its substance is rent both physically and chemically. But in these ways energy is absorbed in the meaning of the principle of dissolution.

Digression for a moment may be permitted to say that the agency of bodies of diverse chemical and physical properties as excitants of coagulation is similar to, if not identical with, 'contact action.' It is well known that solids have the capacity for condensing gases in contact with their surfaces, but the researches of M. Konovaloff intimate a capacity of solids for dissociating gases. 'Platinum enjoys this property to a high degree, but also many other solid bodies, glass among them, the intensity of its contact action obviously depending upon several circumstances ; its chemical composition, the structure of its surface and its temperature, as also upon the density of the gas it is brought in contact with.' [1]

H. Ernst Freund, in ' Wiener medicinische Jahrbucher,' 1886, Heft 1, has shown that blood will not coagulate when drawn into a vessel the inside of which is coated with vaseline ; nor will fibrine separate when the blood is stirred with glass rods if these are smeared with oil. And soaked fish-bladders and parchment-tubes are similarly anti-coagulative. These facts, which I have taken from ' Nature,' No. 852, vol. xxxiii., are readily assimilated with preceding considerations. Coagulation of the blood is not prevented only by ' vital influences,' as we were

[1] *Nature*, vol. xxxiii. p. 350.

wont to believe, but is determined by the quantity and the quality of the molar and molecular energies presented from without.

In seeking to learn if dissolution is further subscribed to in the qualitative changes of function and structure which transpire when blood dies, there is to be marked first that loss of speciality which results from the rending of corpuscles and plasma. Moving for specific ends from one part of the body to another, the living corpuscles and plasma lose, it must be conceded, in definiteness of motion as they become dead fibrine. This must be granted also of the intermolecular motions through which exist the vitality and functions of cells and plasma.

Loss of definiteness of matter, though appreciable, coagulation being a change from speciality to generality, is conspicuous at a later stage of blood dissolution ; that is, when it is decomposed and resolved into imperceptible gases.

Blood in dying also loses in the coherence and heterogeneity of its matter and motion. While living it is an organised system, each part being subservient to the whole and the whole to the parts ; and shows much heterogeneity of function and structure, especially in its formed elements, the white and red cells, blood-plaques, microcytes, &c. But coagulated or dead blood is quite without such co-ordination of forces as would be called function, and its varied histology gives place to uniformity on the destruction of its cellular components.

§ 2. Thrombosis and Embolism.

Thrombosis is a particular case of blood-coagulation, and exhibits as the latter does the essentials of
the dissolutional process. The following description
of artificially-produced thrombosis indicates the nature
and conditions of the phenomenon :

' The mesentery of a frog is exposed and subject
to microscopic examination. A vessel of some size,
an artery or vein, is chosen, and its wall in some way
injured, as by twitching it slightly with forceps, or
by placing a small crystal of common salt near it.
Very soon white blood-corpuscles begin to adhere at
the injured part. As the blood passes over it successive layers of white corpuscles adhere, and a growing
clump of them is formed. Along with the white
corpuscles a stray red one may be insinuated, or there
may be several red ones. The clump so formed, be
it wholly white or partly mixed with red corpuscles,
may be carried off, in which case a new one begins to
form, but the clump may remain fixed and be con-
tinuously enlarged by successive depositions of corpuscles from the circulating blood. In course of time
a change occurs in the appearance of the clump, the
white corpuscles lose their individual outline to a
great extent, and the clump gathers itself together
into a grey granular mass, in which neither by acetic
acid nor by staining are the majority of the white
corpuscles to be discovered. It has, indeed, very much
the characters of fibrine which has been obtained by
whipping the blood outside the body. The clump of

white corpuscles, in fact, by the disintegration of the corpuscles and the attraction of the fibrinogen from the blood plasma, has converted itself into a fibrinous coagulum.'

Thrombosis as here typically portrayed is accompanied, we may safely say, by a disintegration of matter. In an assemblage of leucocytes or blood-plaques (the latter are thought by some observers to be the chief agents in the construction of a thrombus), the protoplasm of the individual cells coheres in spheroidal forms. When thrombosis results, this protoplasm is subjected to radical rearrangement, moving from the former centres of aggregation so that the spheres are obliterated (chap. i. § 2).

The known causes of this morbid change prove in many cases, and imply in others, an absorption of energy from the outer surroundings. It is most plainly shown in thrombosis induced experimentally, as when foreign bodies are introduced into veins. But less so when apparently set up solely by the stagnation of blood in vessels, in the apices of the ventricles, in the auricles, or about the columnæ carneæ. Here thrombi are probably due indirectly to stagnation and directly to the structural and functional alteration to which stagnation gives rise in the endothelium.

When thrombi grow by the accretion of layer upon layer of corpuscles, as in aneurismal coagulation, and as in the extension of thrombi from the uterine to the iliac veins, the surface of the coagulum plays the part of a foreign body in transferring a molecular motion to the circulating corpuscles sufficient to overthrow the balance of their inherent forces and effect their continuous deposition.

Where thrombi form at the mouths of wounded vessels, upon valves that have suffered inflammation, as in endocarditis and phlebitis, upon tumours protruding into vessels, or upon the walls of atheromatous arteries, the conditions sustain the inference that a passage of molecular energy from the immediate investment of the blood is the cause of the coagulation. In these cases, as in cases of inflammation, the *remote* origin of the changes in surrounding structures, upon which thrombi so commonly depend, is not in all instances certainly known. Nevertheless, in numerous instances, the starting-point in the series of causes is indubitably among the forces externally incident. If in endocarditis the qualitative change in the living membrane of the heart is the immediate forerunner of thrombosis, the inflammation of the endocardium we find related to some impurity of the blood, and the cause or causes of this impurity may generally be traced to some form of external energy. This is the case with endocarditis from rheumatism, measles, and scarlet fever.

The remote antecedents of thrombosis sequential to aneurism, phlebitis, and atheroma, may, in part, be traced to the conditions of the individual life. Efficient causes of aneurism are found in atheromatous disease and increased arterial pressure. The frequency of aneurism among those whose occupation entails great stress being put upon the heart and vessels, shows us that the absorption of energy in undue functional exercise is an important factor. Soldiers, ironworkers, and others who perform severe manual labour, are the chief sufferers from aneurism. And atheroma itself may possibly be related to the same causes. Its common seats are those points in the arterial system where friction of the blood is most felt, as at the arch of the aorta, at the bifurcations of arteries, in the pulmonary artery when the right ventricle is hypertrophied. There are also unknown factors in the production of atheroma.

Of phlebitis, its associations with the specific fevers, pyæmia, and gout, may be alluded to as indicating its connection with external incident forces.

The remaining qualities of the dissolutional process can be inferred of thrombosis, since it is but a variety of blood-coagulation in which the change from a definite, coherent heterogeneity to an indefinite, incoherent homogeneity, was perceived, indeed, to obtain.

The softening, putrefaction, and desiccation to which thrombotic coagula are liable exemplify dissolution carried a step further; and so does the so-called organisation of thrombi. In the process of becoming organised the substance of a thrombus is not an active participant; it is dead matter and cannot give origin to a new formation. Its *rôle* is a passive one; the fibrinous substance is disintegrated and resorbed by the wandering cells from neighbouring structures. These cells uniting to form, with young vessels from the *vasa vasorum*, a granulated tissue, a cicatricial mass having the characters of connective tissue is developed.

Particles detached from the substance of a throm-

bus may be carried into different divisions of the
arterial system and, becoming arrested in arteries or
capillaries, give rise to embolism. This involves transformations recognisable as dissolutions. Emboli
have also, of course, other sources of origin, as neoplasms, vegetations, fat, pigment, &c. They may
bring about inflammation and irritation of vessels, and
in the brain the phenomena of apoplexy; but these
results we shall pass over, being here interested only
in the infarctions which embolism occasions.

Infarction is a common consequence of the plug-
ging of end-arteries, and may be either hæmorrhagic,
in which case the wedge-shaped piece of tissue
possesses the general lineaments and attributes of a
blood-clot; or the infarction may be necrotic, in
which case the metamorphosis is a species of coagulation-necrosis; a solid homogeneous mass is substituted for the normal structure. In both of these
effects of an embolus the individualities of the tissue
elements are sooner or later completely effaced by the
redistribution of matter; the mass exhibiting no
marks of organisation.

The conditions under which necrotic infarction
occurs have suggested that stoppage of nutritive
supply is the proximate cause of the death of the
tissue. Can local or systemic death from starvation
be harmonised with the doctrine of dissolution, which
requires an absorption in excess of outside forces?
Dissolution from innutrition implies a deficiency of
ingested energy, since food is the source of organismal power.

The answer is, that, failing due nourishment, the

organism, or any of its units, becomes prey to its circumstances. It is no longer able to resist, from its store of ingrained power, the disintegrating actions that beset it ; the wear and strains of its own substance, mechanical pressures, the chemical and thermal actions that encompass it. So that while withdrawal of food-material is in itself a subtraction from and not an addition to the total energy obtained from without, it makes possible such an absorption as suffices to cause disintegration and dissolution.

One of the immediate resulting phenomena of embolism of an artery is diapedesis. As an element of the dissolutional process diapedesis has received notice in the chapter on Inflammation. The intimate changes in necrotic infarction will be given in the next section.

§ 3. GANGRENE, COAGULATION-NECROSIS, AND ALLIED METAMORPHOSES.

In gangrene of external parts the predominating transformations are those of putrefactive decomposition, which transformations strikingly coincide with the terms of the formula. When dead tissues are freely exposed to the actions of the outer forces there is a copious consumption of kinetic energy, and a proportionately quick disintegration of substance. By invisible motion in the shape of heat, and by the chemical action of the atmosphere, including its living constitutents—putrefactive bacteria and other microorganisms—the tissues may ultimately be reduced to their simplest indefinite inorganic elements. The proportional relation just mentioned between the quantity of absorbed motion and the rapidity of dissolution is seen in that dryness of the air and lowness of temperature (a deficit of liquid, chemical, and thermal motion); the absence of bacteria, and protection from oxidation, retard the process of dissolution.

When necrosed tissues are internal and not exposed to the influences of forces without, the changes differ from those of ordinary decomposition. Then the metamorphosis is known as coagulation-necrosis, and is comparable to blood-coagulation; the solidified tissue corresponding to the fibrine. As the death of white corpuscles is essential to blood-clotting, so is the death of tissue-cells to coagulation-necrosis. The living cells contain, as the leucocytes do, fibrinoplastin and ferment, and the lymphatic fluid which bathes them, fibrinogen. If the cells die, and there is liberation of their fibrinoplastin and ferment, the tissue may coagulate.

Upon the structure of the tissue the effect is similar to the effect of coagulation upon the blood. At first the morphology may be distinguishable; later the nuclei of the cells disappear, then the cells themselves are obscured, and at last the tissue becomes a nearly structureless or sometimes fibrillate substance without differentiation of parts and without organisation.

As just remarked in the preceding section, necrotic infarction is an example of coagulation-necrosis. It occurs also as hyaline or Zenker's degeneration of muscle, where 'to the naked eye a portion of the muscle is seen to be pale and glassy-like, resembling the flesh of uncooked fish.' Muscle which undergoes this transformation loses its transverse striæ, its fibres become friable, and readily break crosswise.

Diphtheritic exudation is another example. The false membrane in diphtheria is partly an inflammatory exudation and partly the coagulated peripheric layers of mucous membrane. Coagulation-necrosis has been observed in paralysed muscles, and *rigor mortis* is probably a similar coagulation of the sarcocele.

Tube-casts of the kidney are sometimes the coagulated epithelium of tubules. Caseous necrosis is a kindred change. 'The necrosed tissue presents the same obscuration of the nucleus as is found in coagulation-necrosis, but in addition all details of structure are completely obscured by the general opaque granular appearance presented by the caseous material.'

Further examples of this variety of retrograde metamorphosis are found in the changes of desiccation and mummification, in softening or colliquefaction, and in sloughing.

In each of the foregoing examples the change is more or less disruptive and destructive of the elements of the tissues. From the displacement of their particles of matter from the former foci of distribution a fine or coarse disintegration results.

Our knowledge of the causes of these degenerations is wanting in completeness, but those causes ascertained imply an absorption of external energy.

The acute fevers, as typhoid and scarlet fever, give rise to hyaline degeneration of muscle. Here, besides febrile heat-energy, there are the chemical energies of the specific poison tending to shake the molecules and particles of the living matter asunder. In all likelihood, with the progress of knowledge the origin of these fevers as related to external conditions will become determinately established, this being quite the drift of gathering facts. The case of diphtheritic necrosis lends itself to the same vaticination.

Tube-casts, whatever their composition, are very probably the result of the injurious action, upon the glandular cells, of chemically perverted or degraded blood and urine, whether arising from disease of the kidney cortex or from the quality of the blood depurated by the renal gland. And changes in the quality of the blood and urine are already, in very many disorders, attributable to forces acting on the organism from its environment (chap. vi. § 2).

Caseous necrosis may be causally connected immediately or mediately with syphilitic poison ; with the conditions giving rise to tuberculosis, prominent among them being the tubercle bacillus.

Desiccation and mummification are invariably associated originally with the incidence of some force or set of forces that has killed the tissues outright; and the details of the metamorphosis are decided by the relations of the dead tissue to its environment.

Colliquefaction, especially of central tissue, is often a consequence of embolism, and that form of softening known as sloughing is a commonly observed consequence of destructive agents. The slough following the application of the actual cautery, the electric current, or caustic potash, will be at once recalled.

And in addition to the disintegration of matter and absorption of motion which these degenerations more or less clearly display, the morphological and physiological changes are quite overtly a transition from relative heterogeneity, definiteness, and coherence to relative homogeneity, indefiniteness, and incoherence. In all, the distinctiveness of the tissues and their elements is lessened, order gives place to disorder, and sameness is substituted for difference.

§ 4. ATROPHY AND THE REMAINING RETROGRADE METAMORPHOSES.

A disintegration of matter is not discernible in the *process* of simple and numerical atrophy; it is, however, implied in the results. The dwindling of a testicle from the pressure of a hydrocele, or of muscle from paralysis, requires us to think that since gross substance has passed into the imperceptible form the particles composing it have been dispersed. The disintegration is generally a molecular one, and therefore invisible. But sensible disintegration occurs in atrophy in the multiplication of nuclei (chap. v. § 2). This is witnessed both with the atrophy of adipose tissue and of muscle. And the integrated fat of adipose tissue undergoes separation into distinct droplets.

All atrophic changes, general or local, are the effects of conditions fundamentally the same. In

E

every case waste is in excess of repair; that is, the
absorption of motion preponderates over its dissipation (chap. i. § 3, p. 16).

The waste may be excessive in relation to the ingested nutriment. Emaciation following stricture of
the œsophagus serves as an example of this. Or,
the food supply being normal, the tissue changes are
greater than normal; this is observed in the wasting
from undue functional activity, as when the testes
atrophy from overuse; and in hæmorrhages, prolonged suppuration, and exhausting diseases.

In senile atrophy the power to assimilate nutriment falls short of wear and tear, and this favours
the absorption of disintegrating energy. Diminished
functional action, in like manner, leads to atrophy by
reducing the demand of the tissues for aliment, and
by consequence lessening their resistance to whatever forces are incident upon them. Atrophy from
mechanical pressure shows a more direct relation to
the absorption of energy. In most cases of atrophy
the causes are not single and simple but multiple
and complexly interconnected; yet it seems probable
that the conditions of any particular case will be
found on investigation to be those required by the
principle of dissolution.

For the remaining correspondences to the formula
we may take extreme examples of atrophy. Then
we find either in tissues or organs a reduction to the
extreme of indefiniteness, incoherence, and homogeneity. Functions and structures utterly disappear, and with them all organisation and differentiation. The atrophy of bone and pulmonary vesicular

emphysema, as instances of partial atrophy, may be profitably compared with the formula.

Fatty infiltration is oftener a physiological than a pathological change, and therefore may be passed over for the changes of fatty metamorphosis. In the latter the substance of tissue-cells is actually converted into fat ; the fat is not directly derived, as in fatty infiltration, from the oleaginous and other constituents of food.

First the cell-body, and later the nucleus, are replaced by granules of fat. In the final stage, with the disappearance of the cell-wall, all that remains is a disintegrating congeries of fatty particles—corpuscles of Gluge. Thus the previous specialised organised unlikeness of structures and functions is represented by simple uniform unorganised fat.

The traits of dissolution are especially well seen in cerebral softening, in which fatty degeneration plays a leading part. ' The white substance of the fibres first coagulates, then breaks up into masses of various sizes (myeline), and these usually undergo more or less fatty metamorphosis. The cells of the neuroglia, the small blood-vessels, and, when the grey matter is implicated, the large nerve-cells, are also involved in the change. The tissue is thus converted into broken-down fibres, granular matter, and molecular fat.' But the process and results are essentially the same in all structures—in muscles, in vessels, and the tissues of viscera.

A satisfactory demonstration of absorptions of

energy as the causes of fatty metamorphosis appears to be impossible as these causes are at present understood. When resulting from phosphorus and alcohol poisoning, or from high temperatures, the absorption is evident. It cannot positively be said to occur when fatty metamorphosis proceeds from deficient nutritive and blood supply, as it does so frequently. The chemical changes involved in the transformation of protoplasm into fat are unknown, though imperfect oxidation is a recognised condition. Undoubtedly the albuminous constituents of the cells split up into the more simple substance, and the decomposition, we must suppose, is effected by incident energies.

Mucoid, colloid, and amyloid degeneration respectively accord in part with the formula. The metamorphosis is of the albuminous elements of the tissues ; these are turned into dead homogeneous material of less complex chemical composition— mucin, colloid and amyloid matter.

When the mutations are advanced, all distinction of parts may vanish, functions are abolished, and no signs may remain of the former varied, concerted, and determinate structural units. As with the degenerations previously considered, the present changes are probably chemical decompositions involving the removal of matter from around its former atomic or molecular centres of arrangement.

Nothing is known as to the etiology of mucoid and colloid decay, and the conditions of lardaceous or amyloid disease are not well understood. The frequent association of the latter with syphilis and long-continued suppuration point to the indirect influence of external forces.

Calcareous and pigmentary degenerations are not pronounced as pathological changes, and so do not exemplify dissolution very distinctly. The deposition of calcareous particles and pigment in the substance of cells does, however, lead to partial obliteration of histological elements, and to inco-ordinations of function.

In the tissue transmutations of pyrexia, 'cloudy swelling,' or granular degeneration, as it is called, the same obscuration of the forms and contents of cells is observed. The nature of the change and its association with acute disease signify its dependence upon the action of disintegrating energies.

§ 5. WALLERIAN DEGENERATION OF NERVE AND ACUTE NEUROTIC ATROPHY OF MUSCLE.

These pathological states are taken together because they have much in common in respect of the microscopic structural changes.

It is desired to show their remarkable exemplification of dissolution.

Wallerian degeneration may be induced by section of a nerve; it is also associated with injury and disease of the central conducting paths of the cord. If in divided nerves the distal portion is not allowed to reunite with the central portion, it will suffer Wallerian paralytic degeneration, with the following microscopic appearances.

Note in how many ways matter is disintegrated.

The substance of the medullary sheath—the myeline—becomes first separated into cylindrical masses, and afterwards completely broken up into globular masses and granules.

And in addition to this physical disintegration there is a chemical one : the myeline is decomposed, fat remaining as the residue. There is also disintegration by proliferation ; the interannular nuclei multiply so that three or four may be seen within the space of an internode. At a late stage the axis-cylinder divides, and is ultimately destroyed ; absorption of this element and of myeline being followed by shrinking of the nerve-sheaths ; and nerve-fibres are then collapsed and wasted.

A new formation of protoplasm within the neurilemma is spoken of, and, if it actually occurs, is a subordinate evolution (chap. iii. § 1). Usually fat granules and myeline are interspersed through the normal protoplasm, and this may give a mere semblance of protoplasmic increase.

Thus, when Wallerian degeneration is set up by nerve-section the absorbed mechanical energy, which disintegrates the nerve by severance of its continuity, initiates a disintegration of all the specialised individual neural elements. And the principle of dissolution is further satisfied by the resulting disarrangements of function and structure ; by the reduction to extreme indefiniteness, for axis-cylinder and medulla pass out of existence ; and by the effacement of all functional and structural elaboration.

Three stages of neurotic muscular atrophy have

been distinguished : simple atrophy, atrophy with division of nuclei, and atrophy with cirrhosis.

Taking the stage marked by the multiplication of nuclei, which serves our purpose the best, we find changes quite analogous to the changes of nerve in Wallerian degeneration. The contractile substance of the fibres assumes a granular degenerated appearance, becomes fatty, and is finally absorbed. The nuclei of the muscle corpuscles multiply abundantly, and sometimes fill the sarcolemma, the muscle-fibres in the end being wholly extinguished.

Subsequently the nuclei of the endomysium, or the corpuscles of connective tissue, proliferate and form new connective tissue. This, called cirrhosis of muscle, is, however, a secondary regenerative process quite opposite to, and not to be confounded with, the process we are contemplating.

Neurotic muscular atrophy is not as definitely related to the action of disintegrating forces as Wallerian degeneration from the simple division of nerves, for its conditions are often very intricate. It is frequently the consequence of injury or disease of the motor ganglion cells of the grey matter of the cord, and is found joined with 'acute central myelitis, spinal apoplexy, fractures and luxations of the vertebral column, infantile spinal paralysis, and related diseases.' [1] In some of these an absorption of energy is determinable, but this cannot be asserted of those cases where the first causes are unknown.

We see how the many contrasted elements of

[1] *A Treatise on the Diseases of the Nervous System*, by James Ross, M.D., LL.D.

muscle-fibre—the sarcous elements, muscle corpuscles. Hensen's and Krause's membranes—are reduced to the grade of uniform granules and fat; and that all order and speciality of both structure and function are brought to an end.

CHAPTER IV.

THE CHANGES INDUCED BY VEGETABLE AND ANIMAL PARASITES AS EXEMPLIFYING DISSOLUTION.

PARASITES, by their presence within or upon the body, excite, for the greater part, diseases of the inflammatory class ; and inflammation, as we have seen (chap. i. §§ 2, 3, 4), accords closely with the formula.

Beginning with vegetable parasites of the natural order Schizomycetes, micrococci play a superordinate part in septicæmia and pyæmia. In these affections the disintegrating action of micro-organisms upon the cells and fluids of the body has been clearly established, and the general. changes of function and structure are those of dissolution (cf. chapters on Special Diseases).

Anthrax, or splenic fever, caused as surely by micro-organisms (bacilli) as any other disease, evidences exactly corresponding changes by hæmorrhages, ecchymoses, inflammatory exudations, suppuration, and sloughing. The efficient condition of scarlet fever is believed to be the inception of a micrococcus. And we see in these, as in other diseases conjectured to emanate from the actions of this kind of parasite, that the disintegrations, if proceeding from the assigned causes, are due to the reception of external motion. For, whether micro-organisms disintegrate the body by the direct influence of their protoplasm upon animal protoplasm, or by the chemical agency of their decomposition products, disruptive energy is communicated.

Bacterium termo is concerned in putrefactive decomposition where protoplasm is split up into peptones, peptones into leucin and tyrosin, and these into simpler substances. In the septicæmia of mice, bacilli may be seen in the act of breaking up the substance of white corpuscles.

The pathogenic moulds, or hypomycetes, are the causes of Favus, Tinea tonsurans, Tinea circinata, and Tinea sycosis, in which the disintegrations and other changes of inflammation are found. In Tinea unguium the nails become fragile, that is, disintegrable.

Of the Entozoa, or internal animal parasites, it will be remembered that Distoma hepaticum produces rot—significant word—in sheep ; and other Trematoda, as Distoma sinense and Distoma hæmatobium, produce paralytic and hæmorrhagic dissolutions.

Would it not uselessly postpone the exposition of diseases to be made in later chapters, it might here be made plain that the primary phenomena of all parasitic diseases are in agreement with the principle of dissolution. The essential functional and structural perturbations from Tænia solium and its cysticercus, from Tænia echinococcus, Bothriocephalus latus, Trichina spiralis, and the rest, are changes to diminished definiteness, coherence, and heterogencity ; and that absorbed energy is the cause of these changes is quite patent. Intercurrently, there are, as in other diseases, minor evolutions. We must so consider the formation of the hyaline membrane found in hydatid cysts, and the new development of encapsuling connective tissue which the presence of certain entozoa leads to.

It is needless to say that the Epizoa, or external animal parasites, Acarus scabei, pediculi, and the larvæ of insects, yield phenomena in strict parity with the foregoing.

CHAPTER V.

NEOPLASMS AS EXEMPLIFYING DISSOLUTION AND EVOLUTION.

§ 1. THE INFECTIVE TUMOURS.

INFECTIVE tumours, or granulomata, occupy, both histologically and clinically, a position intermediate between the inflammatory new formations and the true tumours. Granulomata correspond morphologically to embryonic reparative tissue—granulations—and, like the true tumours, are reproduced in distant places, being infective. Their infectiveness, however, rather resides in their causes—in most examples micro-organisms—not in a power of autonomous growth, such as the true neoplasms exhibit. We may, therefore, take the granulomata for examination before the other forms of new growth.

Since infective tumours manifest a tendency to the formation of cicatricial tissue, the life-history shows alternations of dissolution and evolution.

Syphilitic granuloma may be considered as typifying this class of new growths. If we scrutinise the microscopic features of a hard chancre, there is found beneath the epidermis a replacement of the normal tissue by a congeries of indifferent cell-forms. Now

whence come these ? Are they derived from the
normal units of the organism ? or are they imported
with the virus of syphilis ? That syphilitic granuloma is capable, though with difficulty, of differentiation into connective tissue may be trusted alone as
conclusively testifying to a community of nature between the indifferent cells of infective tumours and
the ordinary formative cells of the body. If the
cytoblasts of the new growth were specific their
developmental changes would be specific.

Concerning the origin of these indifferent cells, it
appears safe to infer that the virus of syphilis sets up
an inflammatory process, destructive in some cases of
the higher structural units. From the more stable
connective-tissue cells, and perhaps emigrant blood-
cells, there is produced a cellular mass with the
characters of granulation tissue. But whatever the
process of the metamorphosis, we are more especially
interested in the results. Without question the functionally and structurally complex specialised and
multiform normal tissue-cells are replaced by simple
cells having little organisation and great uniformity
of parts. And with equal certainty the parts dis-
placed have been destroyed, and therefore their substance has been disintegrated, and this from an
absorption of energy, as the virus of syphilis. Hence
the transformations of matter and motion, culminat-
ing in syphilitic granuloma, are the transformations
of dissolution.

If the indifferent cells, of which granulomata are
chiefly and characteristically composed, pass through
the further changes which end in the production of

connective tissue, the formula of evolution will, as in
the case of repair, generalise those changes.   If the
transformations end, as they often do, in caseation,
coagulation-necrosis, and ulceration, these, as was
seen (chap. i. and chap. iii. § 3), are a continuation
of the process of dissolution.

The morphology of the remaining varieties of
infective tumour—acute miliary tuberculosis and
tubercle, glanders and farcy, lupus, leprosy, and
frambœsia, bovine tuberculosis and actinomycosis—
does not differ in necessary nature from the morphology of syphilitic granuloma ; as in the last-named,
the lesions are essentially a local conversion of normal into granulation tissue, inclosing in addition to
irregular cell-forms those chemical and organic elements which give specificity to the new formations.

Our knowledge of the causes of these diseases,
incomplete as it is, is an implied recognition of
energy transferred from the environment to the organism.   To the *Bacillus tuberculosis* is assigned the
leading part in the genesis of tubercle and perlsucht.
The nature of the agent or agents producing glanders
and farcy is unknown ; these diseases are, however,
highly contagious—a fact suggesting that their causes
may come from without.

Lupus, leprosy, and frambœsia are surmised to
depend each upon a peculiar virus or micro-organism ;
and actinomycosis is etiologically ascribed to the ray-
fungus, whence comes the name of the disease.

§ 2. The Simple and Compound Tissue Tumours (excluding Cystomata and Teratomata).

The phenomena of tumour formation will not fall into like order with the morbid phenomena we have already surveyed, unless the data are true. General laws when applied to new classes of particulars yield new products only when the particulars are purged of inaccuracies. Since, therefore, our knowledge of the nature and causes of tumours is admitted by all to be at present extremely unsettled, it is requisite to attempt an adjustment of some of the data to be now dealt with.

The subject is one of transcendent interest on philosophical as well as on practical grounds, and is deserving of ampler treatment than we shall here be able to accord it. But the object is not to offer a rationale that will be on all points conclusive to the physician. One having such general validity as will justify the application of the principles of evolution and dissolution will satisfy present purposes. To present an argument that might be thought worthy of the subject would be to destroy whatever symmetry this work may possess. As it is, the proposed attempt to co-ordinate the facts will necessitate a divagation of such dimensions that I fear the good-nature of the reader will be strained. Yet some indemnification will accrue if, as is anticipated, the tentative conception to be unfolded ultimately proves to contain the essentials of a true hypothesis. First, we may consider the nature of tumours, and it will afford an introduction to the subject if we look for a moment at the tenability of an hypothesis that has obtained much advocacy—the hypothesis of Embryonic Remains.

Of late renewed attention has been directed to Professor Cohnheim's suggestion that tumours are produced by the formative activity of embryonic rudimental cells. It is supposed that, owing to some development fault, embryonic cells not required in ordinary growth and differentiation are left over in the tissues. Under unknown influences these embryonic remains germinate and form neoplasms.

In speaking of the reasons *pro* the hypothesis, it may first be said that the hypothesis has higher credentials as applied to certain teratomata and other congenital tumours. It will be remembered that these are excluded from the present discussion.

Important evidence is not wanting to prove the existence of the required germs, and there are good reasons for regarding them as the source of mesoblastic neoplasms. Undeveloped pieces of cartilage have been discovered by Professor Virchow in the shafts of long bones. These cartilages are believed to give origin to tumours. Mr. Bland Sutton, F.R.C.S., examining fœtal fingers, has discovered little nodules of undeveloped cartilage in the phalanges, and from these may arise subungual exostoses. Odontomata and dentigerous cysts ' may be positively ascribed to pre-existing teeth-germs which never advanced to maturity.'

As indirect evidence it is argued that cysts are known to originate in obsolete canals and tubules;

as in the tubules of the parovarium and the duct of Gartner. Tumours are frequently found where development is presumed to be complicated, and therefore where errors are like to arise. The orifices of the body are such places ; the mouth, the cardiac and pyloric openings, the external *os uteri* where ' Müller's ducts opened into the urogenital sinus.' At these points there are transitions in the character of the epithelium.

This is, in brief, the foundation upon which Cohnheim's hypothesis rests, though the argument might be strengthened from collateral sources.

Against the hypothesis it has been urged that the proof offered of the existence of embryonic rudimental cells is inadequate. Excepting the cartilage islands nothing is actually known of such cells.

The hypothesis is infertile in explanation, not supplying us with a better understanding of the general pathology of neoplasms. It is also in conflict with many facts ; for instance, it is opposed to the well-established dependence of many varieties of tumour upon the agency of external irritants ; nor would it be valid to account for one form of neoplasm grafted upon another, as happens in the cancerous degeneration of adenomata.[1]

As an alternative hypothesis, to fill if possible the want which the preceding leaves unsatisfied, it is here suggested that the formation of a tumour is a ' seeding' of the natural plastids of the body. As repro-

[1] The writings of Mr. J. F. Payne, M.B., F.R.C.P., in the *British Medical Journal* for 1874, part i., may be consulted for much that bears upon the acceptability of this explanation of the origin of tumours.

F

duction for the entire individual is effected by organs—ovaries and testes—specialised for the purpose, reproduction for the individual units is accomplished by the units themselves. Since the process is by fission or nuclear segmentation, it is an example of agamogenesis, or asexual genesis, the only mode of reproduction known in Protozoa ; and in all probability it is a *reversion to this primal type of multiplication.*

In the course of evolution the transition from the protozoan—simple, unicellular—condition, to the metazoan—compound, multicellular—condition, was marked by the assumption on the part of special cells of the procreative function. Protozoan asexual genesis is, however, retained by the cellular units of Metazoa ; growth and repair are effected by asexual cell-multiplication ; but the reproductive process is not a separation : it is an accrescence. Growth and repair thus differ from the simplest kind of genesis in being a continuous, not discontinuous, increase of cells. It is to be observed that in Protozoa growth and reproduction are one ; the aboriginal form of organic increase is a discontinuous multiplication. Since this point is of some importance as a constituent of the conception now being presented, further exposition is desirable.

In tracing out the evolution of structural forms among the different animal and vegetal groups, Mr. Spencer has argued most forcibly to show that, beginning with aggregates of the first order, which consist of unicellular organisms living and propagating as discrete bodies, the superior orders arise by the union and compounding of these single cells. ' The transition to higher forms begins in a very unobtrusive manner. Among these aggregates of the first order, an approach towards that union of which aggregates of the second order are produced is indicated by mere juxtaposition. Protophytes multiply rapidly, and their rapid multiplication sometimes causes crowding. When instead of floating free on the water they form a thin film on a moist surface, or are embedded in a common matrix of mucus, the mechanical obstacles to dispersion result in a kind of feeble integration vaguely shadowing forth a combined group.' By increasing mass and definiteness this second order of aggregates is perfected, and by coalescence with like aggregates and by physiological divisions of labour organisms of the third order arise, and so on through higher degrees of composition. So that, assuming the primordial mode of organic growth and reproduction to be a separation of the proliferating units, coherent aggregates have been evolved by arrest of this separation, and specialised reproductive organs have arisen with the differentiations wrought by adaptation and natural selection.

The fundamental likeness of nature thus preconceived between growth and genesis is the source of the difficulties met with when attempts are made to construct classifications of forms of growth (increase of mass), forms of development (increase of structure), and forms of genesis (increase of individuals). Classifying from any one of these points of view, there will inevitably be included phenomena classible from the other points of view.[1]

[1] See *Principles of Biology*, chap. vi.

And a like difficulty would confront us in the classifying of tumours as modes of tissue-genesis. If according to the present hypothesis all primary new formations are the asexual products of existing tissue, the offspring are found sometimes closely assimilated to the parent cells both in structure and function—so closely as to be almost indistinguishable, and therefore without that distinctive attribute of a tumour, a more or less definite individuality.

Every degree of autonomy is observable in neoplasms: from the almost completely independent cells of very diffluent round-celled sarcoma to fatty tumours welded with contiguous structures in such manner that no line of demarcation can be drawn between the tissues which are normal and those which are abnormal.

If we make the continuity and discontinuity of the processes a basis of distinction between growth and true genesis, the following table, modified from the 'Principles of Biology,' will give the relations :

$$
\text{Cellular proliferation is}
\begin{cases}
\text{Continuous,}\\ \text{which is } .
\begin{cases}
\text{simple growth}\\ \text{or}\\ \text{metamorphosis}
\end{cases}
\begin{array}{l}\text{including agamogenesis}\\ \text{of plants and animals.}\end{array}\\[2ex]
\text{Discontinuous,}\\ \text{which is } . .
\begin{cases}
\text{homogenesis}\\ \text{or}\\ \text{heterogenesis}
\end{cases}
\begin{array}{l}\text{gamogenesis.}\\ \text{gamogenesis alterna-}\\ \text{ting with discontinu-}\\ \text{ous agamogenesis.}\end{array}
\end{cases}
$$

Continuous asexual reproduction (agamogenesis) is the accretive development of the morphological units of composition in compound plants and animals. Each separate shoot of herbs, shrubs, and trees, consisting of an axis, foliar appendage, and axillary bud, is regarded as an individual ; and in the compound

Hydrozoa, each bud in a colony of corynids is one of a community. This form of reproduction it appears better to distinguish as a form of growth, making the difference between growth and genesis one of un-interruption and interruption respectively of the processes.

We may say, then, that normal growth, hyper-trophy, repair, and so-called inflammatory new forma-tion, to which neoplastic growth bears many resem-blances, are distinguished from true genesis in that they are continuous processes ; and we reach the important fact that neoplastic growth is like genesis in being discontinuous—discontinuous in the sense that the reproduced cells are independent of the re-producing cells. *Both functionally and structurally tumours are segregated with varying degrees of definite-ness from the tissues in which they live,* and this auto-nomy is *sui generis* of them.

Let us now see if this supposed derivation of tumours is substantiated by several classes of facts. Does the embryology of tumours render any confirma-tion ? If tumours begin as offshoots of normal parts ; if they are, as assumed, the yield of tissue-germs or seeds, then in the first period of their life-history they will be germ-like in appearance. This, we find, they are. Probably, whatever the structure of a tumour—whether of epiblastic, hypoblastic, or meso-blastic tissue—in the embryonic stage, the morpho-logy is the same.

Tumours of connective-tissue type—mesoblastic tumours—are known to consist at first of ' indifferent tissue,' accumulations of small spheroids of proto-

plasm not to be discriminated from the cells of granulations ; they have, indeed, *the characters of amœbæ,
organisms of the natural order Protozoa.* 'The first
stage in the process of development consists in the
formation of an embryonic tissue, and this embryonic
tissue subsequently develops into the tissue of which
the new growth is composed.'

That tumours of *epithelial* type are, when embryonic, of like structure, is very probable, though
many observations upon this point are not recorded.
The accompanying figure (fig. 2) shows the advanc-

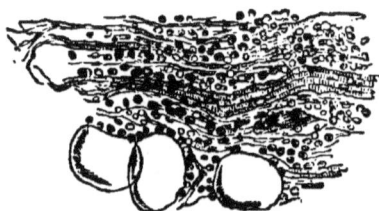

FIG. 2.—SCIRRHUS OF THE MAMMA.

A thin section from the most external portion of the tumour, showing the
smal.l-celled infiltration ('indifferent tissue') of the muscular fibres and
adipose tissue in the neighbourhood of the gland. × 200.

ing edge of a carcinoma ; and we may, I think, believe
that its 'indifferent tissue' represents an early stage
in the development of all epithelial tumours.   The
almost immediately following illustration of nephritic
cancer greatly encourages this belief.

Starting thus from normal cells, the growth of a
neoplasm, its mere increase of mass, appears to result
partly from the multiplication of its own units, and
partly from the continued division of the protoplasm
of healthy tissue.   In primary cancer of the kidney
this germination of the body's substance has been

actually witnessed. Then 'the tumour is virtually
a cancerous degeneration of the organ. The kidney
may be completely converted into a tumour which
sometimes attains a very large size, retaining the
general shape of the organ, and covered by its capsule.
But in some cases only a part of the kidney is involved, and in that case while the affected part retains
the general shape of the organ, although enlarged,
the remaining piece of the kidney has quite the
normal appearance. To the naked eye it is as if a
portion of the kidney were transformed, and with the
microscope it can be seen, at the margin of normal
and pathological, that the tumour is advancing by
a conversion of the proper kidney tissue. *The epithelium of the tubules is multiplying so as to form the
cancerous epithelium*, and is becoming irregular in
form *while the cancerous stroma is being formed of the
connective tissue of the organ.*' Also, 'at the marginal
parts of an epithelioma of the tongue, for instance,
it can nearly always be seen how the normal epithelium is extending and penetrating inwards.'

And it consists with the present hypothesis that
not only normal but abnormal tissues are the parents
of new growths. If neoplasms are the asexual progeny of the individual cells of the organism, and the
genetic process is a reversion to the simplest form
of reproduction, then pathological structures may
exhibit the phenomenon. Now various kinds of
imperfectly developed and redundant tissue are well
known to be the sources of neoplasms. From nævi,
warts, and moles they not infrequently arise. And
uterine polypi, dermoid and other tumours, may

undergo cancerous degeneration, a tumour giving
origin to a tumour.

It will readily be granted that if the cells of these
eccentric structures are subjected, like normal cells,
to the conditions necessary for neoplastic activity, such
agamogenesis may be manifested by them. Presently
it will be shown that their conditions of life are peculiarly favourable. With this view it can also be
understood that ' cartilage islands,' teeth-germs, and
functionless structures anywhere, have the potency
for producing tumours.

Another group of harmonising facts is comprised
in the histology of tumours. The morphological relations among tumours of the connective-tissue class,
and among those, too, simulating the epithelial tissues, are just what might be inferred if neoplasms
are local offshoots of existing parts. ' Every pathological growth has its physiological prototype.' We
see the parents in the progeny.

Of tumours of mesoblastic tissue, if the germ-cells
reach full development we have the reproduction of
adult tissues—fibroma, myxoma, lipoma, chondroma,
osteoma, odontoma, lymphoma, angioma, myoma,
and neuroma. If development does not advance beyond the embryonic stage there results round-celled
and melanotic sarcoma ; and spindle-celled, alveolar,
and myeloid sarcomas are slightly higher forms of the
preceding. Likewise, in epithelial neoplasms according as the seeds develop there are encephaloid and
scirrhus cancers with attempts at glandular arrangements of the elements ; or adenomata with sometimes

perfect gland-structure. And so with epitheliomas, cylindrical and papillary; they repeat the parent structures. If we consider the developmental changes in tumours as illustrating the formula of evolution, these important relations will be brought out in explicit detail.

Round-celled sarcoma, regarded as the most embryonic of the connective-tissue tumours, is often in point of development but little in advance of the feebly integrated homogeneous protoplasmic particles of which, as we have found, all neoplasms of this class at first consist. 'It is an exceedingly soft tumour, often half-diffluent.' Its cells are clustered with little connecting substance; nor does round-celled sarcoma generally possess that compactness conferred by the development of a capsule. By this as well as by the superior kinds of sarcoma the formula of evolution appears to be illustrated. In the process of structural differentiation the small, round, discrete plastids of the embryonic tumour become relatively less numerous, cellular, and nucleated; a more definite intercellular substance appears; this commences as a soft amorphous or granular matter, and later becomes fibrillate, the entire structure acquiring greater concreteness.

In the gliomatous, melanotic, spindle-celled, and myeloid varieties of sarcoma, the elements show further advance in integration, speciality, complexity, and multiformity. There is, with more or less regularity, a progressive firmness of consistence, the cells lie nearer together, and the growth is often encapsuled. And with the assumption of fusiform shapes, the appearance of nucleoli, and the new arrangement of the cells, there may be deciphered a growing resemblance to the diversiform, definite, and co-ordinate tissue whence presumably they were derived.

Through the myxomata and lipomata and their sub-varieties, teaching the same truths, we reach neoplasms copying mature forms of tissue. Biologically and histologically speaking, the distinction between the sarcomata and fibrous, cartilaginous, and osseous tumours is principally one of development, however wide their clinical differences. Beginning like the sarcomata in germ-protoplasm, these new growths, in the nearness of their approach in structural characters to full-grown tissues, describe by their life-changes yet more vividly the traits of evolution. And so of tumours of higher orders of tissue —the lymphomata, angiomata, myomata, and neuromata.

Passing to neoplasms originating in epithelial tissue, we find the counterpart of round-celled sarcoma in encephaloid cancer; a new growth remarkable for the want of union among its elements and for the simplicity of its structure. 'Encephaloid cancer is of soft brain-like consistence.' 'The proportion of its stroma is very small.'

Thence the transitions of growths leading on the one hand to scirrhus and adenoma, and on the other to epithelioma and papilloma, accord with the formula of evolution. The advance in integration and the concomitant dissipation of motion—the advance from the general to the special, from the simple to the complex, and from the uniform to the multiform—is very distinguishable.

In scirrhus the growth is often hard and compact ; the cells are comparatively few, and the stroma abundant. Sometimes, as in mammary cancer, there are well-formed acini, the neoplasm closely repeating the morphology of the parent glands.

Glandular tumours frequently resemble precisely the tissue of glands. An adenoma of the mamma is reported as containing acini so advanced in organisation as to produce milk.

Epithelioma, reproducing in different degrees the peculiarities of the epithelial tissue of the regions which are its common seats, exhibits by individual development and the divisions of its class the same degree of correspondence to the principle of evolution. Flat-celled epithelioma of the mouth, pharynx, vagina, &c., is described in various stages of development of cells and stroma ; it often grows outwards, as does villous cancer, and approximates to the papillomata. And cylinder-celled epithelioma, as of the stomach and intestines, sometimes partakes of the characters of cylinder-celled glands.

The family relationships indicated by the transitional varieties of neoplastic growth have been given perspicuous expression by Mr. Frederic S. Eve, F.R.C.S.[1] 'The glandular structure of cancer is most beautifully shown in cancer of the thyroid gland and of the large intestine. The morbid growth in the thyroid consists of gland-tubes lined with one or more

---

[1] The *British Medical Journal*, Feb. 17, 1886, p. 299.

layers of regularly arranged columnar or spheroidal
epithelial cells . . . if the existence of transitional
forms can be proved between the innocent and malignant types of the epithelial-tissue tumours, these are
as potent as any anatomical fact can be in the connective-tissue series, whether of the hard or soft parts.
Fibrous tumours occur in every variety, from the
dense compact nodules composed of parallel bundles
of well-formed fibrous tissue to the soft and sometimes semi-gelatinous growths, which, however, are
not malignant.' Descending from the fibro-sarcomata,
' we reach the spindle-celled and finally the round-
celled varieties of sarcoma, in which the elemental
round-cell makes not the slightest effort in the direction of organisation, not even to the formation of a
spindle-cell.' The osteomata are said by Mr. Eve also
to present analogous intermediate forms, and the fact
is noted that tumours of epithelial origin may have
the primitive characters of round-celled sarcoma.

All these facts are at one with the supposition
that neoplasms result from tissue agamogenesis. It
quite accords, if the various tissues reproduce their
kinds, that there should be such individual and
generic differences among tumours as pathologists
have recorded. The continuation of internal forces
in the germs is expended ' in working out a structure
in equilibrium with the forces to which ancestral
organisms were exposed,' modified by existing conditions.

It may now be inquired whether the circumstances
under which neoplasms are known to originate also

bear witness to their being the products of a generative process. This will involve an investigation of the causes of tumours, a subject of superlative importance.

To appreciate the bearing and force of much that will follow, the essential nature of genesis should be comprehended; it should be conceived as a process of disintegration. We may take the case of tumour-genesis. Tumours being hypothetically the descendants of existing or pre-existing tissues, we have to look at the transition from the normal to the abnormal type of structure. The formed protoplasm of the body gives birth, by a partition of its substance, to particles capable of undergoing autonomous development. Now what essentially is this parturient process? It is a *disintegration* of the living matter of the tissues, and in so far a dissolution of it.

The process of cell-proliferation has already been alluded to (chap. i. § 2) as a species of disintegration; but to thoroughly comprehend it as such, there must be noted its analogue in those primitive organic types which prefigure for us the remote ancestors of the superior types.

Referring to previously cited cases of self-division among the Diatomaceæ, Mr. Spencer observes : [1] ' In these cases the multiplication is so carried on that the parent is lost in the offspring—the old individuality disappears either in the swarms of zoospores it dissolves into, or in the two or four new individuals simultaneously produced by fission.' And 'among the Protozoa as among the Protophyta, there occurs that

[1] *Principles of Biology,* vol. ii. p. 427, E. 1867.

process by which the individuality of the parent is wholly lost in producing offspring, the breaking up of the parental mass into a number of germs. An example is supplied by the lowest class, the Gregarinæ. This creature, which is nothing more than a minute spheroidal nucleated mass of protoplasm having a structureless outer layer denser than the rest, but being without a mouth or other organ, resolves itself into a multitude of still more minute masses, which when set free by bursting of the envelope shortly become amœbaform, and, severally assuming the structure of the parent, go through the same course.' Also, of creatures multiplying sexually by the formation of sporangia and by conjugation, we read that 'as the entire contents of the parent-cells unite to form the sporangium the individualities are lost in the germs of a new generation. In these minute simple types sexual propagation just as completely sacrifices the life of the parent or parents as does that form of asexual propagation in which the endochrome resolves itself into zoospores.'

In these aboriginal modes of reproduction in which the dissolution of the mass is complete, and in those higher modes throughout the animal and vegetable kingdoms where only portions of the parturient mass are sacrificed, the common relation of likeness is the *disintegration* of the parent body. In the primordial types, complete disintegration; in the derived types, partial disintegration. Disintegration is, as we know, a mark of dissolution.

And now we may inquire as to the causes of this disintegration. Under what conditions are tumours generated? There have been submitted cogent reasons for supposing that tumour-formation is a form of genesis. It may be expected, then, to show some conformity to the laws of normal genesis. We shall again find it useful to state in advance the general conclusions that have been reached.

The conditions of genesis at large, and the known circumstances under which tumours arise, lend a high degree of certainty to the inference that tumour-germs are set free by the tissues in the process of dying. The histogenic dissolution or disintegration by proliferation, which has just been considered as belonging to tumour genesis, is the expression of a nearly exhausted tissue-vitality. The phenomenon may be put in the same category as the multiplication of nuclei accompanying the death of cells from degenerative changes, e.g. the atrophy of muscle, the degenerations of nerve or of the cells of glands. And it is paralleled by certain productive phenomena of the inferior types in the organic world. It will presently be seen that various kinds of exhausting influences acting under particular conditions are the causes of tumours. We may first look at some of the conditions of normal genesis, and afterwards at the conditions of tumour genesis.

In plants there appears to be a relation between reproduction and relative innutrition. As this relation, if real, may prove to have important significations, we may quote at some length the observations upon it made by Mr. Spencer.[1]

'The relation between fructification and innutri-

---

[1] *Principles of Biology*, vol. i. chap. vii.

tion among plants was long ago asserted by a German biologist—by Wolff, I am told. When some years ago I met with the assertion I was not acquainted with the evidence on which it rested. Since that time, however, I have, when occasion favoured, examined into the facts for myself. The result has been a conviction, strengthened by every further inquiry, that such a relation exists. Uniaxial plants begin to produce their lateral following axes only after the main axis has developed the great mass of its leaves and is showing its diminished nutrition by smaller leaves or stunted internodes. In multiaxial plants, two, three, or more generations of leaf-bearing axes or sexless individuals are produced before any seed-bearing individuals show themselves. When, after the first stage of rapid growth and agamogenetic multiplication, some gamogenetic individuals arise, they do so where the nutrition is least; not on the main axis or on the secondary axes, but on the axes that are most removed from the channels which supply nutriment. Again, a flowering axis is commonly less bulky than the others: either much shorter, or, if long, much thinner. And further, it is an axis of which the terminal internodes are undeveloped. The foliar organs, which instead of becoming leaves become sepals and petals and stamens, follow each other in close succession, instead of being separated by portions of still-growing axis. Another group of evidences meets us when we observe the variations of fruit-bearing that accompany any variations of nutrition in the plant regarded as a whole. Besides finding, as above, that gamogenesis commences only when the luxuriance of early growth has been some-what checked by the extension of the remoter parts of the plant to some distance from the roots, we find that gamogenesis is induced at an earlier stage than usual by checking the nutrition. Trees are made to fruit while still quite small by cutting their roots, or by putting them in pots ; and luxuriant branches which have had the flow of sap into them diminished by what gardeners call ' ringing,' begin to produce flower-shoots instead of leaf-shoots. Moreover, it is to be remarked that trees which, by flowering early in the year, seem to show a direct relation between gamogenesis and increasing nutrition, really do the reverse ; for in such trees the flowering buds are formed in the autumn—that structure which deter-mines these buds into sexual individuals is given when the nutrition is declining. Conversely, very high nutrition in plants prevents or arrests gamogenesis.'

These facts, with others of the same import, are brought forward by Mr. Spencer as evidence that high nutrition is one of the conditions of agamo-genesis and low nutrition of gamogenesis. Here we appropriate them simply to show that discontinuous reproduction is conditioned in some cases by rela-tively low nutrition.

In further support of this we also read [1] of plants —the cryptogamia—that ' fissiparously-multiplying cells in some cases fly asunder, while in other cases they unite into threads or laminæ or masses ; and fronds originating proliferously from other fronds, sometimes when mature disconnect themselves from

[1] *Principles of Biology*, p. 48.

their parents and sometimes continue attached to them. Whether they do or do not part is clearly determined by their nutrition. . . . That disunion is really a consequence of relative innutrition, and union a consequence of relative nutrition, is clear *a posteriori*. On the one hand, the separation of the new individuals, whether in germs or as developed aggregates, is a decaying away of the connecting tissue ; and this implies that the connecting tissue has ceased to perform its function as a channel of nutriment.'

I have myself recently observed an interesting corroboration of this. Bacilli cultivated upon hard-boiled egg—a very rich soil—may acquire gigantic proportions. *Bacillus anthracis* thus propagated I have seen of extraordinary thickness, and many times longer than those grown in gelatine.

Reproduction by spores in micro-organisms is also seemingly related to a deficiency of nutritive pabulum. In the culture of bacilli it is found that when the gelatine or other culture-medium is becoming exhausted the bacilli are resolved into spores. With test-tube cultures, spore formation takes place from above downwards as the food is used up ; on the surface of the gelatine where the nutriment has been consumed spores are abundant, but in the deeper parts multiplication is proceeding by simple division of the bacillary rods. This phenomenon of spore-formation, it should be stated, is viewed by Dr. Klein, F.R.S., as the effect of oxygen at the surface of the gelatine. Here is called to mind the observations of Dr. Downes and Mr. T. P. Blunt, M.A., on the effect of light and oxygen upon the protoplasm of bacteria. From the results of their experiments, which I take from Mr. Spencer's 'Factors of Organic Evolution,' these observers conclude that light and oxygen may combine for the destruction of the protoplasm of bacteria, as they do for the destruction of inorganic matter. May it not be that spores are made up of partly oxidised protoplasm, and that by this reduction to a semi-inorganic constitution there is conferred that resistance to heat and other forces which so excites our wonder in microbic spores? These germs would then, perhaps, represent by their molecular structure one of the evolutionary stages marking the passage from the not-living inorganic matter to individualised protoplasm.

But to continue the argument. Some features of the segmentation of the ovum appear to me to have a kindred meaning ; division of the germ-mass is de-cided, other things being equal, by the distribution of the food-yolk—the nutritive aliment of the pro-toplasm of the germ. That this is so is not uni-versally assented to ; but it has the high sanction of Professor Balfour, from whose work, 'A Treatise on Comparative Embryology,' the following facts are extracted :—'Most of the eggs, which have a per-fectly regular segmentation, are of a very insignificant size and rarely contain much food-yolk : in the vast majority of eggs there is present, however, a con-siderable bulk of food material, usually in the form of highly refracting yolk spherules. These yolk spherules lie embedded in the protoplasm of the ovum, but are in most instances not distributed uniformly, being less closely packed and smaller at one pole of the ovum than elsewhere. Where the spherules are fewest the active protoplasm is necessarily most concentrated ; and we can lay down as a general law that the velocity of segmentation in any part of the ovum is, roughly speaking, proportional to the concentration of the protoplasm there ; and that the size of the segments is inversely proportional to the concentration of the protoplasm. Thus the segments produced from that part of an egg where the yolk spherules are most bulky and where, therefore, the protoplasm is least concentrated, are larger than the remaining segments, and their formation proceeds more slowly.' This effect of the food-yolk upon the division of the protoplasm is shown by what is called unequal segmentation. Here the disposition of the viteline substance is unsymmetrical, the larger spherules being at the lower and the smaller at the upper pole. After the primary furrows have been formed, the segmentation proceeds in suchwise that in the later stages there are 128 segments in the upper half of the ovum and 32 in the lower. This action of the food-yolk is extensively manifested, and on the ground of it a special terminology is applied. Ova which segment uniformly are termed ' alecithal,' ' as implying that they are without food-yolk, or what food-yolk there is is distributed uniformly. Telolecithal ova are ova in which the food-yolk is not distributed uniformly, but is concentrated at one pole of the ovum. When only a moderate quantity of food-yolk is present, the pole at which it is concentrated merely segments more slowly than the opposite pole ; but when food-yolk is present in very large quantity, the part of the ovum in which it is located is incapable of segmentation and forms a special appendage known as the yolk-sack.' In centro-lecithal ova the food-material is centrally disposed ; the segmentation involves only the circumferential zones and advances centripetally until the food-yolk is reached, leaving a central undivided portion.[1]

It must be remembered that the segmentation changes of an ovum are not to be identified with the subsequent embryonic changes, the differentiation of the epiblast and hypoblast, but with the primordial mode of growth in Protozoa, which is a discontinuous multiplication of cells (p. 66). Balfour observes : ' Although the various types of segmentation which have been described present very different aspects, they must, nevertheless, be looked on as manifestations of the same inherited tendency to division, which differ only according to the conditions under which the tendency displays itself. This tendency is probably to be regarded as the embryological repetition of that phase in the evolution of the Metazoa which constitutes the transition from the Protozoon to the Metazoon condition.'

In what way food-yolk influences segmentation Professor Balfour does not vouchsafe explanation. Obviously, as he suggests, where food-yolk is present almost to the exclusion of protoplasm these segmentations will not occur, the quantity of protoplasm being too small. He believes that cell-division is the result of molecular changes affecting the cohesion

[1] *Op. cit.* vol. i. p. 91 ; *vide* fig. 48 (2).

of the protoplasm—' alternations of cohesion are
produced by a series of molecular changes the external indications of which are to be found in the
visible alterations of the body of the cell and of the
nucleus prior to division.' I think the facts quoted
favour the conjecture that food-yolk so influences the
play of molecular forces in the protoplasm of an
ovum as to give relative cohesion to its particles, and
that this influence is nutritive. But, whatever the
true interpretation, there is clearly, from the previously given examples, a causal connection between
genesis and relatively low nutrition.

Genesis is also related to high nutrition. That
animals and plants are fertile or infertile according
as they are respectively well or ill nourished, is almost
a commonplace truth. Therefore it must be that
the just observed connection of reproduction in the
lower forms of life with deficiency of nourishment
is a relation to *relative* innutrition. In the case of
plants, fructification takes place at those terminal
points where nutritive changes are least active; but
unless the general nutrition reaches a certain level
reproduction will not set in at all. When once
inaugurated, it will, too, be furthered by abundant
nourishment. 'Were it otherwise, the manuring of
fields that are to bear seed-crops would be not simply
useless but injurious. Were it otherwise, dunging
the roots of a fruit-tree would in all cases be impolitic; instead of being impolitic only when the
growth of sexless axes is still luxuriant. Were it
otherwise, a tree which has borne a heavy crop

should, by the consequent depletion, be led to bear a still heavier crop next year ; whereas it is apt to be wholly or partially barren next year—has to recover a state of tolerably high nutrition before its sexual genesis again becomes large.' Among animals also this is evident ; a good supply of nourishment facilitates reproduction, and the casting off of new individuals is generally a casting off of a surplus—of something over and above what is required for individuation.[1]

There remains to be mentioned the antagonism between genesis and growth, development, and expenditure. By the laws of reproduction, whatever energy is used up for growth of mass, for advance of structure, and for functional work, is so much deducted from the energy available for reproductive processes. It has been established by induction, and is required by deduction, that there exists an inverse relation, variously complicated, between genesis and growth, development, and expenditure. For the elaborate arguments by which these laws have been founded the reader is referred to the 'Principles of Biology,' vol. ii., part vi.

And now we have to learn whether the conditions of tumour-genesis are coincident with these conditions of normal genesis. Are the relations just considered reproduced in essential features in the phenomena presented by neoplasms ? In the first place, does the production of morbid growths evince any relation to local nutrition ? The answer is in the

[1] Vide *Principles of Biology*, vol. ii. chap. ix.

affirmative ; but the relation is traversed by other relations in such manner that only in a general way is the truth recognisable.

The obstruction to be overcome by the blood-propelling forces increases at a high rate with each increment of distance from the central organ (the heart) of the nutriment-distributing system. Hence the heart, and organs clustered round the great arterial trunks, will, if other things are equal, be more favourably situated for obtaining nourishment than will distant parts. Now as we proceed from the trunk to the periphery of the vascular tree there is found a general increase in the liability of organs and tissues to new growths. (We must leave out of consideration all secondary tumours, for their distribution is usually determined by the distributing systems, the blood and lymph-vascular systems.) Tumours of the heart itself are 'exceedingly rare,' and practically this organ is exempt from any liability to primary new formations—a singularly impressive fact. Tumours of the lungs, central organs, are almost invariably secondary. ' Cancer of the lung is a disease of which I have no knowledge.' [1] Primary cancer in them is not described.

In the very vascular liver, tumours are oftener seen than in the lungs ; but, omitting cavernous angioma, primary growths are comparatively seldom found in the hepatic substance. And they are very infrequently observed in the spleen, but as we pass to the pancreas and kidney they become more common.

[1] Dr. Wilks, F.R.S.

Receding in the same direction still further from the centre of the vascular system, we reach the bladder, in which papilloma and cancer occur with frequency.

Of the organs of generation in the male the testicle is very sparingly nourished. The spermatic arteries are long and slender vessels, and arise from the abdominal aorta almost as high as the point of origin of the renal arteries. Correlatively with this, and not fortuitously, we may think, these glands are peculiarly disposed to generate morbid growths. Sarcomas and cancers are the usual forms, and less commonly enchondromas, fibromas, myomas, &c.

Turning to the organs of generation in the female, the ovaries, analogous to the testes, and similarly circumstanced in respect of their blood-supply, are similarly ready to undergo neoplastic degeneration. In the uterus, myomas are so common that, it is said, one woman in every ten is afflicted ; and sometimes the new centres of development are so numerous that 'there may be as many as fifty attached to the same uterus.' The frequency of cancerous disease in this organ is notorious.

Passing upwards from the heart, leaving for the present the entire alimentary tract, we have remarked the rarity of tumours of the lung, which organs are in immediate contiguity with the heart. Now the bronchus and trachea enjoy a similar immunity, but the distantly situated larynx gives origin to many kinds of morbid growth, including papilloma, fibroma, mucous polypi, sarcoma, and epithelioma. The favourite seats, too, are the projecting vocal cords and epiglottis. The brain is a fertile soil for new growths.

This general proneness to tumour-genesis betrayed by tissues according as they are near to or far removed from the centre of the nutrient system appears in another way.

We may roughly regard the blood-vessels of the entire organism, and of its organic or visceral parts, as having central and peripheral dispositions. Thus, for the whole body there are the great central arterial trunks ; the general external periphery, the skin ; and the general internal (involuted) periphery, the lining tissues of the alimentary canal. For the separate organs there are usually a fundus, body, or base—as of the uterus, stomach, and mamma—and extremities and surfaces. Now these terminal parts, whether surfaces or points, are generally *at the ends of the vascular branches* ; and it is remarkable that here neoplasms, with few exceptions, take origin.

Looking at the case of the first of these divisions, how marked is the proclivity of the skin and alimentary canal to generate tumours, and how comparatively seldom are they generated by the interperipheric tissues !

Then, of the viscera individually, the gastric and duodenal tissues furnish a very large number of cancers. In this the stomach is excluded from the rule that organs near the heart show exemption. The usual starting-point, however, of the cancerous growths is the submucous tissue of the *pyloric* and *œsophageal orifices*, though they are commonly observed clinically along the lesser curvature. Cancers do not often grow from the fundus, nor other tumours from the peritoneal surfaces where the stems of the vessels are.

Apparently a similar relation to vascular supply is traceable with tumours of the intestine. In the cæcum, sigmoid flexure, and rectum, cancers are observed with a gradually increasing frequency as the terminus of the gut is approached. But in the centrally-situated ileum and jejunum they are scarcely ever found.

And if we examine the case of the uterus, or of the mammary gland, similar facts are pressed upon us. It is chiefly at the peripheries of the visceral systems of vessels, at the orifices and upon the internal surface of the uterus and near the external surface and at the nipple of the mamma, that neoplasms first appear.

These illustrations might be multiplied considerably by noting the common situations of tumours of the mouth, nose, ears, eyes, brain, and penis. From the gums spring various forms of cystic tumour; and from the edges of the lips and tip of the tongue, epithelioma. In the brain the peripheric hemispheres are the favourite sites for glioma. Cancers spring generally from the *dura mater* and *pia mater*; sarcomata, from the 'free surfaces of the interstitial spaces.' Epithelioma of the penis begins usually in the glans penis. Further, in the bones, speaking generally, osteo-sarcoma arises at the extremities of long bones, and from the external and internal peripheries, the medulla and periosteum.

Very striking is the fact that sarcoma has a predilection for those parts of the bones where growth last ceases, i.e., at the ends opposite to the line of direction of the nutrient arteries. The lower epiphysis of the femur is the last to join the shaft, and the nutrient arteries are oppositely directed—towards the upper epiphysis. Of seventeen central tumours of the femur, ten were situated at the lower epiphysis. The upper epiphysis of the tibia and humerus join the shafts at twenty-five and twenty years respectively, when growth in the rest of the bone is complete; and the nutrient arteries are directed downwards in both bones. Of thirteen tumours of the tibia, 'eleven started in the upper epiphysis; and of eight affecting the humerus, all but one were in the upper epiphysis.' The fingers and toes are extremely liable to enchondromata, and upon the ends of divided nerves it is common to find small fibro-neuromata.

Although the relation is confused by conditions we have purposely refrained from taking into account, but which will presently be dwelt upon, it is undeniable that there is an increasing propensity to generate tumours in tissues according as they are remote from the source of nutritive supply. And this may be regarded as analogous to the connection with relative innutrition which is observed in the seeding of plants.

Tumour-formation is also, doubtless, related to high nutrition and excess of food. There are not at present any facts to show what this relation is, but I strongly suspect, on several grounds, that it is one of very great practical importance.

Can there be discerned of tumour-genesis the antagonism to growth, development, and expenditure established for normal genesis ? It does appear to be feebly shown.

First to be perceived is the broad fact, having many exceptions, that the tumour-developing period of life is the reproductive period. It is not during the time of most active growth and development, but in the interval between puberty and senility, that tumours are of most common occurrence. Adenomata are 'almost invariably observed between puberty and the age of thirty years.' Cystic adenoma is ' most common between seventeen and forty,' and myo-fibromata of the uterus have never been seen before puberty. Of sixty-three cases of central tumours of bones, only two occurred before the age of sixteen. Of fifty-six cases of sub-periosteal tumours, four-fifths occurred between the ages of sixteen and forty. Again, cancer of the breast and uterus are most common between the ages of forty and fifty years, the period immediately preceding senility of these organs. ' These diseases are very rare before thirty.' Cancer of the tongue, mouth, and œsophagus is most frequently observed between the ages of fifty and seventy.

That other coincidences with the laws of genesis are not observable is perhaps to be expected. If the phenomenon of neoplasm-formation is a reversion to Protozoan reproduction the conditions will only partially resemble the specialised conditions of Metazoan reproduction. But from the considerations that have been advanced it will, I think, be safe to conclude that *the genesis of tumours is a phenomenon of the same order as reproduction in general.*

Yet this will not suffice for a satisfactory application of the principle of dissolution. It will be
necessary to pass under notice, with all convenient
brevity, the remaining factors in the production of
neoplasms.

The reader who has followed the preceding paragraphs will have mentally passed the criticism that
the usual sites of new growths, besides being at the
vascular peripheries, are also points where tissue irritation is particularly concentrated ; and that such
irritation is probably a considerable factor in the
generation of tumours. Unquestionably this is so.
Irritation must be recognised among the important
co-operating conditions. Chemical and mechanical
irritants are positively known to be causes of neoplasms. Chimney-sweeper's and coal-miner's cancer
and the scrotal epithelioma of persons exposed to the
vapours of naphthaline are those specially referred to.
And among these may be included forms of papilloma
—condylomata, venereal and other verrucæ—resulting from the irritation of unhealthy secretions.

It is not improbable that kindred causes are at
work in the production of gastric cancer. Usually
the overt declaration of this disease is preceded by a
long period in which dyspepsia is suffered, when the
prominent pyloric and œsophageal orifices will be
acted upon by the hurtful products of imperfect
digestion. And in the cæcum and rectum, if fæces
are unhealthy their frequent stagnation at these situations will occasion irritation of the intestinal mucous
membrane.

Leucorrhœal and morbid menstrual discharges

are possibly adjuvant causes of uterine cancer ; and epithelioma of the lips and tongue is known to be determined by the action of tobacco or the irritation of clay pipes—smoker's cancer. Here may be pertinently mentioned the lines of cancerous nodules noticed in the tracks of sutures after operations for the removal of carcinoma.

It would not require prolonged search to find further evidence of causes of this nature ; but we must pass to another class of actions known to be concerned in the bringing forth of tumours.

In the cases of so-called ' traumatic malignancy ' a definite relation has been made out to the infliction of mechanical injury. Sarcoma following the fracture of bones is not seldom met with. There is next that comprehensive class of cases which recognises inflammation as one of the antecedent conditions. of new growths. Eczema of the nipple as the precursor of duct cancer, to which attention has been called by Sir James Paget, is one among many possible examples.

Cases evidencing another kind of instrumentality are those showing the relation of tumour-genesis to imperfections of development or structural inferiority of tissue. Reparative or cicatricial tissue often undergoes neoplastic degeneration. Tumours have many times been observed to emanate from the ear-ring cicatrices of the aural lobes, and with special frequency from the cicatrices of burns. ' It is not uncommon to see a cancerous growth in the base of a healed ulcer.' Possibly traumatic neoplasms grow from the newly-formed tissue of repair.

Then, as already adverted to, tumours themselves

germinate : polypus of the uterus will become
cancerous, that is, will cast off cells of less developmental power than the parent growth. Uterine
myomas and mammary adenomas originate respectively secondary fibromas and cancers. An identical
phenomenon is the production of tumours by nævi,
warts, and moles.

The relics of once functional structures are disposed to neoplastic activity. In the next section it
will be seen that they are the common seats of cystomata. The pineal gland, now regarded as the
' vestige ' of an ancestral median eye, sometimes becomes hyperplasic, forming a considerable tumour ;
and hyperplasia of the peripheral Pacchionian bodies
is recorded. From the walls of cysts formed in
obsolete canals neoplasms may grow.

These are the various conditions under which new
growths are known to arise. What is the uniformity
amidst their diversity ? What is there in the functionless state, in relative innutrition, in the continued
action of irritants, that is common as conditioning
the genesis of tumours ? With the understanding
that what follows is not presented as a completely.
reasoned theory, but as a rough though in the main
true synthesis, to round off the conception our purposes have demanded, there may be framed a pro-
visional statement of the causation of tumours which
will be comprehensive of all these and other factors.

The data brought forward sustain the conclusion
that the genesis of neoplasms is another example of
histogenic dissolution. The phenomenon must be
classed with the multiplication of nuclei in Wallerian

degeneration and atrophy of muscle, with inflammatory cell-proliferation, with certain of the reproductive changes seen in tuberculous and other infective
tumours, and in parenchymatous degenerations, like
multiple neuritis and nephritis. And in all these
cases the reproductive process is a reversion to that
simplest form of genesis found in the Protozoan
ancestors of the Metazoa, where cells are resolved
into a fresh generation, or give off by division portions of their protoplasm for the making of new individuals. An invariable feature of this simple genesis
is *the drawing to a close of developmental or vital changes.*
The decline of cell-life, we have learnt, is brought
about in many ways. With plants and micro-organisms relatively low nutrition will determine the
division into germs ; at those points on the axis of
plants where germs are produced the sap reaches with
difficulty, and at those points, consequently, the processes of life first diminish. In the case of the simpler
micro-organisms, genesis, we may rightly suppose, is
still more under the empire of directly-acting agents,
and less subservient to inherited tendencies than in
plants. Hence, just prior to the death of the protoplasm from failing nutriment separation into new
centres may be wrought chiefly by the environment.

Inflammatory proliferation is initiated by the
devitalising influence of phlogogenic agents ; the
muscle nuclei sometimes found surrounding *Trichina
spiralis* lodged in the contractile substance are the
product of an inherited proclivity incited into action
by the disruptive actions of the parasite. In like
manner, the increased nuclei of Wallerian degeneration

and atrophy of muscle are liberated in the process of dying, which process the causes of these degenerations set up in the nerve or muscle. Similarly, whatever the cause of multiple neuritis, or nephritis, in which diseases centres are set free as nuclei, one of the effects of this cause is to stop those changes which constitute the lives of cells, and as the life is ending there is a dissolution into germs.

The conditions of tumour-genesis, though peculiar, do not differ essentially. It has been pointed out that peripheric parts, parts furthest removed from the heart or from the vascular centre of the particular organ, are liable greatly beyond others to give origin to tumours. Now just at such places the tissues may be expected first to ripen; besides the feebleness of nutritive changes where blood-supply is relatively deficient, surfaces and extremities are most exposed to surrounding forces capable of bringing on local tissue senility. It is observed that long-continued and exhausting influences are effective instigators of neoplasms; instances are the cases of warts induced by working with the hands in foreign matter, and cases of cancer from the irritation of soot and coal-dust.

The skin and the mucous membrane of the alimentary canal, the most common seats of neoplasms, are at the extremities of the vascular branches, but furthermore are in direct and constant converse with the energies of the outer world. Thus the life-conditions of those units of the body the readiest to procreate tumours are conditions highly favourable for the premature exhaustion of the energies cognised as vital. The molecular motion through which protoplasm undergoes the adjustments to incident forces constituting evolution must soonest run out where the draughts upon it are greatest, and where the store of motion is from the beginning relatively small, and augmented by nutritive substances with relative difficulty.

Why is reparative tissue inclined to histogenic dissolution ? We may answer with the assumption that its capacity for passing through the changes of continued life will be less than that of the tissues from which it was derived, and, playing a more passive part, as cicatricial tissue usually does, it is correspondingly less vascular. Cicatricial tissue is tissue of relatively low vitality. This is true also of germinating tumours ; and in moles, ' vestiges,' and embryonic remnants no demand is made by function for active nutrition ; hence, slight forces impinging, and in obedience to a disposition inherited from remotest ancestors, they disintegrate by proliferation.

But a condition necessary and peculiar to neoplastic growth remains to be noticed. When generative dissolution takes place in ordinary inflammation and degeneration, the germs do not, in general, subsequently appear as tumours. Why is this ? They perish by further disintegration or are absorbed by the surrounding tissues. In the growth of tumours there must be recognised a struggle for existence as going on between the old tissues and their young, and the survival of the latter constitutes neoplastic power. It is easily perceived that the circumstances favourable for tumour-genesis are circumstances inimical to tissue-resistance. At the vascular peripheries, and in functionless and regenerated parts, decline of tissue-life is promoted by the conditions we have noted ; *but this very senescence permits the new growth to gain the mastery.* The tissues germinate, the forces of life being nearly spent; but the new generation of cells is endowed—by what process we know not—with the prepotency of all germs ; and the latter survive in the competition for nourishment. It is usually otherwise in the tissue-germinations of inflammation and degeneration. Here the young cells may be released at any point, in the substance of the heart or on the outer covering of the body ; but the genetic process is set up in tissues whose stores of vitality are not depleted, apart from the influence upon them of the forces causing inflammation. The action of the phlogogenic or other force is limited to the area of its distribution, and the potentialities for continued life in the contiguous tissues may be normal.

If the cells proliferated in inflammation eventually become tumours, the combination of requisitions for tumour-growth have been accidentally fulfilled.

But whether the young cells outlive the old ones will probably be governed by still other conditions : the power of multiplication in the germs, and the systemic health of the person that bears them. In proportion as the seeds have imperfect developmental powers they multiply rapidly, and may survive by reason of their numerical strength. And lowly-organised tumour-cells are usually discrete, readily gaining access to the lymphatic and blood-vessels, through which they reach the general system. Clinically, this is spoken of as malignancy in tumours.

That the general health of the host has much to do with the survival of the parasite is a reasonable inference. Tumours being insubordinate in the general working of the body, and united with it only by the common nutrient system, if the 'tide of life' is low the organism will retire the more easily before the invasions of the parasite. This will be found assimilable with many clinical facts, as those which show that depressing nervous and other influences make patients an easy prey to cancerous disease, and that improved hygienic surroundings will cause a new formation like lupus to disappear. I have now in mind an instance of this, reported, I believe, by Mr. Jonathan Hutchinson, F.R.S. And thus is afforded a means of explaining the casual spontaneous disappearance or diminution of cancers. Cancer 'may dry up and be in fact cured, which is sometimes seen in the breast, and occasionally in the internal organs.' Reproduction by fission is a supersession of special functions by a more general function. Loss of the higher co-ordinations of cell-protoplasm revokes the subordination of the lower. Simple procreation in cells is normally kept in check by the more specialised activities; hence, whatever maintains the integrity of the latter will tend to the subjection of the former.

Benignancy is usually the correlative of superior development in tumours. Here the power in the germs to attain the higher orders of structure presumably arises from a less degree of that agedness of tissue which, by the hypothesis, favours the genesis of tumours. And it agrees that in these cases the mother-tissues oppose the growth of the germs by the organisation of a capsule. Also in harmony is the fact that tumours of higher tissue-type appear earlier in the life of the individual than do the embryonic cancers. Embryonic sarcomas are, however, often seen in early life.

We have spoken of reversion to the simplest form of reproduction as the manifestation of an inherited trait called out by the inception of disintegrating motion. Can there be divined any reasons for such inherited predisposition ? All attributes transmitted by descent were at some period of evolutional history acquired by the interactions of organism and environment, not even excepting the inherited 'spontaneous' variations which form the 'purchase' of natural selection. What, then, in most general terms, were those interactions which initiated or resulted in multiplication by cell-division ? Only very speculative opinion can at present be fashioned upon a question so abstruse ; yet there is a convergence of testimony giving reasonableness to the following view.

Every protoplasmic body, like every aggregate of such bodies, dissipates in the course of life its store of 'vital' (molecular) motion, and in doing so loses that plasticity of molecular structure which permits adjustments to surrounding circumstances. It follows, therefore, that while it becomes in one sense more rigid, it becomes more liable to have its particles separated in coarse disintegration by whatever forces fall upon it. If, then, during the evolution of ancestral Protozoa the molecular disruptions just preceding death—which is always succeeded by coarse disintegration—took the shape of generative dissolution, this would be perpetuated by heredity. Highly conserving the continuation of the species, the tendency to dissolve into germs before life ceased altogether would be appropriated by natural selection, and become deeply stamped in the constitution of the units of compound organisms. Hence, with the recurrence in a general way of the conditions of its evolution, it is natural that there should be such recurrences of the phenomenon as pathology widely illustrates. The evolution of the higher types of the reproductive process seen in the multicellular organisms, its distinction as a function of specialised organs and ministered to by high nutrition, would follow according to well-known principles.

Thus, I think, is made less incomprehensible the apparent nonconformity to rules of morbid action displayed by tumours. The uncertainty of their appearance in a given case is explicable when it is seen that a *concurrence* of circumstances is required. There must be a conspiration of both local and systemic conditions, using the last term in the broadest sense. The presence of one cause of neoplasms will not suffice in the absence of another or of others.

It ensues from all that has been said that a primary tumour is both a local and systemic disease, but that local conditions largely predominate as determining factors.

There are groups of miscellaneous facts not yet noticed to be placed side by side with the hypo-

thesis. It is inexpedient to enter upon their consideration in this work, though the questions they
suggest may possess much immediate economic importance. We must content ourselves with the mere
mention of some of them :—The special causes of
uterine myoma—what connection has this variety
of neoplasm with functional exercise of the sexual
organs? The conditions of neoplasms in infancy
and youth. The relation of neoplastic degeneration
to over-feeding and superfluous body-energy ; to the
ingestion of particular kinds of food. The facts of
secondary growths (metastasis)—are they at one or
at issue with the conception put forth ? &c.

And now, after all this preparation, can it be said,
that dissolution and evolution find further fulfilments in the phenomena of tumour-genesis ? The
evidence of tissue-disintegration has been adduced.

Is the principle of dissolution complied with in
that an absorption of energy causes this disintegration ? The evidence on this head has also been presented. It is clear that traumatic injury, mechanical
and chemical irritants, the intercourse with external
powers which exhausts the means of tissue-life, testify
to inordinate inceptions of motion as the true causes
of the tissue disintegration.

And, unmistakably, the rearrangements of matter
and motion involved are those of dissolution. In
the beginning all tumours, we may not unwarrant-
ably believe, are congregations of lowly organised protoplasmic particles—indifferent cells. If, then, we
compare this relatively simple, general, and uniform

structure, having like functions, with the relatively complex, specialised, and heterogeneous parent-tissue, there truly comes into view the exemplification of dissolution we have sought.

The later transformations through which the germs pass in the process of development, conform, we have found, to evolution, and it remains to refer to the retrogressive changes. Those met with in tumours are of the same kind as the degenerations of normal tissues, and as such they have already been considered. Therefore, all that is required here is their enumeration. They are calcification, caseation, and softening ; colloid, mucoid, and fatty degeneration. Tumours are also liable to hæmorrhage, ulceration, suppuration, and decomposition.

The morbid affection set going in the organism by the expansion of a new growth within its substance may be an inflammation or irritation followed by the development of cicatricial tissue in surrounding parts ; it may be an extension of the neoplastic change to normal tissues from the influence of the tumour-cells ; or it may be a reproduction of the growth in distant places due to the dissemination of these cells by lymphatic, blood-vascular, and other channels. From what has gone before it will be seen at once that all these changes are comprehended by the principles of evolution and dissolution.

Neoplasms interfere with the functions of organs. A tumour in the spinal cord will give rise to disorders of motion and sensation ; in the stomach, to disorders of digestion ; in the bladder, to disorders of vesical function. And there is the cancerous cachexia, the

components of which are systemic and recondite alterations in the physiological processes of the body.
As seen from the point of view of the principle of
dissolution, this class of disturbances consequential to
the diseases under consideration may be more conveniently instanced in succeeding chapters, especially
since the pathology of cancerous cachexia is too
indefinitely understood to admit of separate treatment.

§ 3. The Cystomata, Teratomata, and Malformations.

The very many modes of origin of cystomas, and ·
our ignorance in manifold cases of the conditions of
their formation, quite preclude such references to the
formulas as we might rest content with. Glancing
first at retention cysts, those due to the confine-
ment of normal secretions within pre-existing cavities,
doubt arises whether this disease can truly be called
a dissolution. Sebaceous cysts—wens, for example—
are not obviously accompanied by disintegrations of
matter; rather is matter integrated. This is strikingly
the case of the curious horns growing from sebaceous
tumours. It seems possible to effect a reconciliation
with the formula of dissolution only by regarding the
contents of the cavities as the accumulated products
of normal processes of dissolution. The epithelial
scales, fatty matter and cholesterine of sebaceous cysts,
are excrementitious elements collecting from closure
of the excretory ducts; and the horns just alluded
to are nothing more than exfoliated epidermal cells
dried and hardened by exposure to the air.

It coincides with dissolution that the disease is sometimes caused by inflammation and excessive activity of the sebaceous glands. Unhealthy matters finding outlets through these channels provoke hypersecretion and inflammatory closure of the cutaneous openings. Congenital piliferous cysts, which resemble sebaceous tumours but are congenital and have fully developed hairs growing from them, rather belong to the order of neoplasms.

Cysts arising from the retained secretions of mucous glands, as the glands of Bartholin and Cowper, and also galactoceles, probably originate in the same way as sebaceous tumours. Ranulæ, from occlusion of the ducts of salivary glands, are oftentimes connected with the impaction of a calculus, and with the irritation of bits of vegetable fibre and fruit-seeds derived from articles of food. New growths, calculi, and parasites are among the causes of pancreatic ranulæ, biliary cysts, and hydronephrosis.

The cystic kidney of chronic nephritis, where there is ectasia and dilatation of urinary tubules, must be referred, remotely, to the causes of the latter disease.[1] Congenital cystic kidneys are now thought by some observers to depend upon the degeneration of persistent fatal structures, e.g., the Wolffian body, and not upon the congenital absence of the pelves of the kidneys.

Dilatation of the vermiform appendix, leading to a cystic tumour of this structure, commonly results from closure by foreign bodies or inflammation at the

---

[1] See Part II. chap. v. § 2.

opening into the cæcum. The causation of cysts of the vitelline duct and urachus is very obscure.

In this form of cystic disease, then, the indications of dissolution are exceedingly weak ; only here and there can we detect an initial disintegration of matter and concomitant absorption of motion. Perhaps by an examination of each example in all possible detail a less incomplete correspondence might be effected. But, by the formation of these tumours, we may legitimately say that the organism sustains a loss in the federation, expressness, and variety of its functions and structures.

Exudation cysts call forth similar comments. In these there is excessive secretion in cavities not provided with an excretory duct. The slighter forms of certain examples (bursæ and ganglia) are not diseases, but adaptations of structures to simple increase of function. The enlargement of bursæ and ganglia is the result of increased action of tendons and muscles. When the actions are excessive, and other conditions are favourable, inflammatory dissolution may ensue, which in numerous familiar examples is the clear effect of communicated motion. Ganglia of the extensor tendons of the hands in pianists, the swelled bursa patella of housemaids, and miner's elbow are cases in point.

Most cystomata connected with the brain, spinal cord, and their membranes are cases of malformations by arrest, and will presently be mentioned with these affections. Naming them here, they are anencephalia, meningocele, encephalocele, syringo-myelus, and the varieties of spina bifida.

Many of the cysts in structural relation with the organs of reproduction rehearse the truth we had occasion to note in the chapter on new growths, namely, that 'obsolescent remnants' are eminently disposed to neoplastic degeneration. Relics of the mesonephros, or Wolffian body—an ancestral excretory organ—persist in the hilum of the human ovary ; the tubules of the mesonephros within the folds of the broad ligament are known as the parovarium. It is a credible presumption that these relics are the sources of what are spoken of as ovarian papillary and parovarian cysts, and, in the male, as encysted hydrocele.

The duct of the mesonephros—Gartner's duct— opening into the vagina, is almost certainly concerned in cystic formations. Mr. Bland Sutton has brought forward an impressive array of facts to show that numerous cystic growths of the ovarian parenchyma start in the arrested normal decay of ovarian follicles. There is hardly any doubt that one of the conditions of the genesis of true tumours obtains with these cystic new formations in diseased parts. And it seems as likely, from the observations and reasoning of Dr. W. O. Priestley, F.R.C.P., LL.D.,[1] upon the genesis of cystic chorion—the so-called hydated mole —that cysts of this class are true tumours ; that they originate from asexual reproduction of normal cells ; and that, some of the new generation undergoing retrogressive changes which bring them to the fluid

---

[1] The Lumleian Lectures on the Pathology of Intra-Uterine Death. *British Medical Journal*, No. 1371.

state, the product is the pro-cellular albuminous or gelatinous substance which these cysts contain.

Very little that is positive can be said concerning the causation of cysts of the broad ligament not connected with the parovarium. Cystomata of the Fallopian tube, known as hydro- and pyo-salpinx, and hydroceles of the spermatic cord and tunica vaginalis, may be more satisfactorily brought under the formula of dissolution, causally linked as they are with inflammations often ascribable to external conditions.

Cysts formed by hernial protrusions of intestinal, pharyngeal, and other membranes, by the softening of new growths, by extravasations of blood and serum, and by parasites, clearly begin in the majority of cases as disintegrations of matter caused by the absorption of energy, and are accompanied by the qualitative changes of function and structure belonging to dissolution.

The congenital cervical hygromata call to mind the air-chambers beneath the deep cervical fascia in some apes; but this does not help us to an assimilation of these obscure cases with the principle of dissolution. Certain of the hygromata are classed among the Teratomata, to which we may now pass.

Teratomata are congenital tumours peculiar in the multiformity of their constituent tissues. 'Some of them contain rudimentary skeletal elements, such as those of the spine, pelvis, &c.; as well as rudiments of various organs, like the brain, intestine, different glands, kidneys, and muscles. Others contain tissue-formations of various kinds, such as muscular tissue, cartilage, skin, bony substance, gland-tissue, cysts, &c.; but no definitely formed masses which can be regarded as representing any special organ or member.'

If we might accept the conclusion of Mr. Sutton,[1] that teratomata arise where the three germinal layers, epiblast, hypoblast, and mesoblast are in contact, then these new growths could be classed with the true tumours. They might be regarded as resulting from germination of epiblastic, hypoblastic, and mesoblastic tissues, which would explain their multifarious structure. But beyond a relationship to ancestral vestiges there are no data bearing upon the actual factors concerned in their genesis. Mr. Sutton's investigations tend to show that teratomata connected with the sacrum, pituitary fossa, pharynx, ovaries, and testes, and the lingual teratomata, all take origin in desuete and embryonic structures, as the post-anal gut, Wolffian bodies, lingual canal, &c. It would sustain in a remarkable manner the argument of the preceding section if tissues incapable of further development should thus so widely assume the simple, general, archaic function of agamic reproduction. But since in teratomata the actual determinants of the reproductive process are unknown, it is impossible to harmonise the phenomena with that term of the formula of dissolution which comprehends causes. In a remarkable paper—'Teratomata: an Ætiological Study'—published in the 'Journal of Comparative Medicine and Surgery' for October 1887, Mr. Sutton brings forth evidence, to me convincing, that many congenital dermoids are related to general conditions

[1] *An Introduction to General Pathology.*

of embryonic development. He points out that his conclusions strongly uphold Professor Cohnheim's hypothesis of embryonic remains, as the hypothesis covers this class of tumours. It appears to me that they are not against the theory I have sketched in the preceding section. Therefore, in order to add a word of comment, I take the liberty of inserting Mr. Sutton's introduction to his consideration of sequestration cysts :

'In order to comprehend in its full significance the ætiology of this interesting group, we must consider briefly the main features connected with the formation of the body-walls. In a transverse section of the body of a mammalian embryo at an early period of gestation, we find that it consists of two bilateral portions united by an isthmus. This isthmus contains the notochord, flanked on each side by the mesoblastic somites. Its lateral parts present a smaller dorsal moiety, which may be considered as upgrowths from the larger ventral portions. The dorsal upgrowths tend gradually to fuse across the median line to complete the spinal cord and the skin covering it. . . . Ventrally each half subdivides the inner layer ; the splanchnopleure eventually coalesces with its fellow to form the walls of the gut ; the outer, which is continuous at first with the amnion, also fuses with its fellow to form the body-walls ; hence it is termed the somatopleure. Coalescence may fail to occur at any spot along this extensive line, or even throughout the whole length of it. If it fails at certain spots we may get such conditions as ectopia cordis and eventration.

'Dorsally this line of coalescence extends from the occipital protuberance to the coccyx, then passes beneath the notochord, and extends ventrally, involving the perineum, scrotum in the male, abdominal and thoracic walls, neck, symphysis pubis, and finally becomes arrested at the septum nasi. The line of coalescence also extends from the septum nasi through the mid-line of the palate to the uvula. It also involves the tongue ; the original cleft in this situation extends from the mental symphysis to the hyoid bone. In any part of this line epiblastic involutions occur, which may manifest themselves as dermoids.'

Dermoids appearing in this tract Mr. Sutton arranges in three groups : the dorsal, ventral, and cranial ; and he refers to the fact that the tumours rarely spring from the trunk or central divisions. From the scrotum, sometimes from the penis, and from the neck, palate, and cranium, dermoids commonly grow ; but not on the line of coalescence from the coccyx to the occiput behind, and from the penis to the neck in front. Now it may be, as Mr. Sutton strongly inclines to think, that these teratomata 'originate in detached masses of embryonic tissues.' The line of coalescence is caused by the infoldings of the embryonic tissues ; and along this formative tract the structures are liable, as we know, to imperfections of development. It is therefore conceivable that where growth and development are defective redundant cells may be included and become the germs of tumours. Is it not, however, equally conceivable, and as probable, that from the influence of injurious forces

I

acting during embryonic life the tissues upon the
formative tract revert to agamogenesis, as we have
supposed adult tissues do in the case of non-congenital tumours ?   The tissues upon the line of coalescence are analogous to the growing extremities
of the fruit-bearing axis of a plant, and to the peripheric parts from which cancer and other neoplasms
so frequently originate.   If this explication, which
comprehends rather than runs counter to the one
favoured by Mr. Sutton, is valid, it accounts in a
manner for the relative frequency of dermoids at the
extremities of the line, the scrotum, head, palate, and
neck. Here, perhaps, would first be felt any failure in
the capacity of the embryo for continuous growth, and
centres of discontinuous growth might arise.   The
very different and mixed constituents of dermoids
would be determined by the stage of embryonic development at which the division into tumour germs
happened.

If with some authors we regard teratomata as
forms of acardiac monsters (*acardiaci amorphi*), then
some force which cleaves the developing embryo is
the aboriginal cause (see following paragraphs).

It will be seen that if cystomata and teratomata
arise initially from local disintegrations and give rise
by their remote effects to secondary changes of this
nature, the growths and development of the component tissues, including fibrous capsules, are intercurrent evolutions.

There is to be included under the present heading

a brief closing allusion to the phenomena of malformations.

For the most part in the diseases thus far referred to we have looked for primary causes to the external forces which act upon the organism during extra-uterine life. In the diseases now to be contemplated we look for the first disturbers of function and structure to the external forces acting during intra-uterine life; and notwithstanding the remoteness of the ætiological factors from direct observation, we are not debarred from an attempt to gather the facts together under the principle of dissolution.

Malformations may be referred to external interferences with normal growth and development *in utero*. These interferences may occur at any stage—in the ovum, in the embryo, or in the fœtus; and if in ancestors, then the defect may be transmitted by inheritance, as seen in the cases of supernumerary fingers. Or the causes of the defect may be repeated in the uterine lives of successive or alternate generations with a feigning of the action of heredity.

Cases of malformation by arrest (*monstra per defectum*) are ascribed to actions which stop or pervert the maturative changes of a normal embryo. Among them are classed ‘defective development or disease of the membranes and placenta (uterine or fœtal), adhesion of the amnion to the fœtus, abnormally small quantity of *liquor amnii*, tumours of the uterus, concussions of the uterus with separation of the membranes, hæmorrhages in the membranes or in their neighbourhood, &c. As concerns the fœtus itself, its development may be disturbed by inflammatory affections which it acquires by transmission from the mother (smallpox, scarlatina, endocarditis), or by inherited disease, such as syphilis. In the earlier stages of development abnormal twists and flexures of the embryo may be enough to cause very serious hindrances to growth.'[1] In whichever of these ways deformity is induced, there is no impropriety in saying that absorbed motion is the cause. Be the case of malformation by arrest acrania or anencephalia, where the vault of the skull or the brain is wanting, the causes conjectured are undue pressure by the head-fold of the amnion or acute cranial flexure of the embryo, and in such manner force is incepted as mechanical motion. And Dareste supposes that synophthalmia, where the orbital cavities coalesce and parts of the brain are absent, may be attributed to imperfect development of the head-fold of the amnion. This presumably permits pressure upon the anterior cerebral vesicles. ' In some cases the cause of cleft palate has been found to be a morbid adhesion of the amnion to the face.' Many other forms of malformation receive no specific explanation of their causation ; among these are hernia cerebri and spina bifida, cleft-face and bronchial fistula, clefts of the abdominal and thoracic walls. Defects in the development of the limbs (amelus, peromelus, &c.) are commonly accounted for by supposing ' constriction, and even amputation, by folds or bands of the fœtal membranes or by loops of the umbilical cord.'

[1] *A Text-book of Pathological Anatomy and Pathogenesis*, by Ernst Ziegler. Translated by Donald Macalister, M.A., M.D., F.R.C.P.

The causes assigned, then, of malformations by arrest are such as connote an absorption of motion and disintegration of matter ; and the effects are obviously in most cases a reduction of functions and structures in definiteness, coherence, and heterogeneity.

Double monsters (*monstra duplicia*) and malformations by multiplication of parts are assumed to proceed from complete or partial cleavage of the embryo and its parts. If the cleavage is entire and occurs at a very early stage of embryonic life, e.g., in the germinal area, there result thoracopagi, xiphopagi, &c. ; and should the development of the segments be unequal, there may ensue forms of parasitic twin—fœtus papyraceus, acardiacus, thoracopagus parasiticus, inclusio fœtalis, &c. Partial cleavage of the rudimental embryo may lead to varieties of duplicitas anterior and duplicitas posterior. Homologous triplets are initiated by threefold cleavage ; and cleavage affecting the rudiments of individual structures occasions multiplication of the limbs, mammary glands, &c.

This theory of the ætiology of double and triple monsters is not opposed to the principle of dissolution. There is implied by such cleavage of the embryonic elements the incidence of a separative or disintegrative force either in the individual, or, where the deformity is inherited, in ancestors. The functional and structural changes induced in the organism correspond but vaguely with the remaining clause of the formula. It is said that double monsters and other malformations have been produced experimentally, as by varnishing hen's eggs before incubation. But the active agents in the production of monsters in the ordinary way are wholly unknown.

Of cases of congenital hypertrophy, where members of the body are unduly large, it appears at present impossible to make a satisfactory synthesis.

PART II.

SPECIAL DISEASE.

CHAPTER I.

§ 1. PRELIMINARY : INTERSTITIAL AND PARENCHY-MATOUS INFLAMMATION.

Corollary I.—Incident forces being equally distributed among the elements of an organic system, the dissolution of functions and structures follows the opposite order of their evolution ; the higher and accessory are less stable than the lower and fundamental.

The structural plan of the organs is that of an active proper substance or parenchyma fixed in a passive supporting substance or interstitial tissue. When the process of inflammation terminates in growth of the supporting framework, it is the custom to speak of this as chronic inflammation. Frequently it is called interstitial inflammation, as distinguished from inflammation of the parenchyma. Manifestations of the phenomenon are widely extended in pathology. An increase of connective substance is the constant and conspicuous change in fibroid induration of the heart, fibrosis of the lungs, cirrhosis of the liver and kidney, sclerosis of the spinal cord, and many other pathological states. Commonly, if not always, there is an associated degeneration or disappearance of parenchymatous elements. The adjoined figure, representing the effects of chronic myocarditis, shows the muscular bundles wasted and degenerated, and their places occupied by fibroid tissue. Similarly, chronic inflammation of the liver, kidney, or other organ is found to be accompanied· by decay and atrophy of the parenchyma, or proper substance of the viscus, and by a new growth of interstitial substance. Now the usual and received theory states that

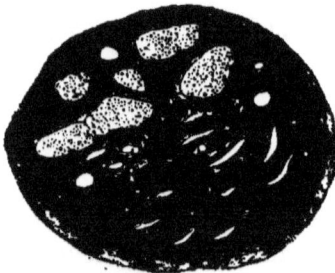

FIG. 3.—FIBROID INDURATION OF THE HEART. A section from the left ventricle. The fibroid tissue surrounds the individual muscular fibres, which are undergoing fatty degeneration. × 200.

the overgrowth of interstitial material is a direct effect of its inflammation, and destruction of the parenchyma is an indirect effect, following the inroad upon it and contraction of the new connective tissue. The connective tissue supports the nutrient vessels, and by its shrinking shuts off the blood-supply of the parenchyma and causes its decay.

There appear to be just grounds for dissenting from this interpretation. All observation leads us to affirm that inflammation is in its very nature depressant and destructive, and not constructive. When we noticed inflammation as illustrative of dissolution we saw that the process is essentially in its primary aspects a disintegration (chap. i. § 2). How, then, can the growth (integration) of connective tissue be an in-

flammatory process ?   I think it is preferable to view
these cases of connective-tissue formation joined with
atrophy and decay of parenchyma, as instances of
overgrowth from removal of the normal resistances
to growth. In a structure that has reached a state of
equilibrium in respect of its tendencies to increase in
mass, there is among the components a balanced
opposition to disproportionate increase of any one or
any group of them. Facts will presently appear
which show that, in a great preponderance of cases,
in chronic inflammation the elements first affected
are the most differentiated and specialised elements—
those constituting the parenchyma. The damage or
destruction of these is not caused by the encroachment
and condensation of the connective tissue, but by the
original disease-producing agent, whatever that may
happen to have been, and *the formative action of the interstitial tissue is in response to the lessened resistance to its
growth which follows injury to the parenchyma.* When
we come to consider the causes of chronic hepatitis and
nephritis—cirrhosis of the liver and kidney—we shall
find that the noxæ are distributed to the proper working substance of the organ, not to the interstitial
substance. Looked at in this way, cirrhosis, fibrosis,
and sclerosis are tied by common nature to repair by
scar tissue, which is essentially a new production of
connective tissue, along lines where the opposition to
growth has been diminished by a destruction of parts.

In the following chapters this conception will be
invoked to aid in making clearer the pathology of
both chronic and acute inflammation of special organs. Acute inflammation, as familiarly understood, is

nearly co-extensive with parenchymatous inflammation ; the damage to parenchymatous elements is the leading feature, and the growth of connective tissue insignificant or absent. Here we shall suppose that acute and chronic disease are related to the intensity, duration, and distribution of the incident forces. Acute disease being usually the correlative of severe, and often sudden, action of irritants, the parenchyma is generally damaged or destroyed without sequent formative changes in the connective tissue. But chronic disease being the correlative of relatively slight and gradual actions, the parenchyma is more slowly affected and time is allowed for interstitial growth.

If we make requisition also of the supplementary doctrine to dissolution (cor. i.), it will be seen that when the morbific forces, in both acute and chronic disease, fall equally upon all the elements of an organ or tissue, the later evolved functional and structural attributes are less resistant than those antecedently evolved. The specialised parenchyma is injured or obliterated before the general basis substance, and the elements of the parenchyma suffer more or less according as they are more or less specialised. The chief reason of this relative instability of the least organised and most specialised elements is that these are the active workers in the body physiological, and are in consequence, largely by means of the blood, brought into closest contact with injurious agents. They are also situated at the expanding distant parts of the nutrient system of the organ. Already we have met with a noteworthy example of this general truth in the facts of true tumour-formation. We saw in the last chapter that the fundamental organs are scarcely known as the seats of primary new growths, but that accessory organs and structures are highly susceptible to tumours. The heart and lungs enjoy singular freedom ; but the sexual organs, the mammary glands, the vocal cords, the epiglottis, the tonsils, and the pyloric valve are among those parts which suffer frequently.

The two principles of interpretation now enunciated will be found, I think, to greatly transform our conceptions of the diseases whose consideration we are to enter upon in the succeeding chapters. Not only shall we find unlooked-for harmonies in the results of morbid actions, but a much greater simplicity in the processes than is generally suspected ; and this last is of immediate practical moment.

We shall now select from the principal nosological divisions further illustrations of dissolution and evolution. Some of the general morbid changes considered in the foregoing chapters will in the following chapters be met with again, though under different aspects. And, as with general disease so with the diseases of organs, the changes regarded in their entirety will be seen to be analytical or dissolutional, minor syntheses or evolutions always being accompaniments.

It is conceived to be important to sustain by induction from accumulated facts the proposition that dissolution proceeds from the absorption of external energy—that diseases are the outcome of environing

conditions ; hence this will be given precedence over
other features of dissolution.   Where the phenomena
of familiar affections can be exhibited in new and
interesting relations by regarding them from the dissolutional point of view, such affections will have
claims upon our attention.   If we followed the established order of nosological tables we should begin
with general diseases—those that affect the entire
system, as the infectious diseases.   But expediency
must be allowed to overrule the demands of usage or
logic ; and in the next section we shall look at some
disorders of the urinary system.

§ 2. ACUTE AND CHRONIC BRIGHT'S DISEASE.

The Structural Changes.—Nephritis, whether of the
acute, waxy, or cirrhotic variety, presents us with
lesions that are clearly changes towards a less definite,
coherent, and heterogeneous structure (chap. i. § 4).
Of the acute form we read that in the first stage 'the
convoluted tubules often present a swollen opaque
appearance, and occasionally contain blood.   On microscopic examination the congestion of the vessels becomes very apparent, and the tubules are found to
be dark and opaque ; their lumen is frequently oc-
cluded.   The individual epithelial cells are granular
and in a state of cloudy swelling.'   In the second or
stage of fatty transformation 'the convoluted tubules
are in many parts occupied by sebum-like material,
and sometimes the straight tubes present the same
appearance.'   Microscopically, some of the tubules
'present under low powers a black appearance due

to fatty degeneration of the contents of the tubules.' Many of them 'are completely blocked up by this material; and sometimes, in making the section, there is such an amount of oil set free, that it permeates the whole structure of the organ and is liable to produce the impression that the fatty degeneration is universal.' In that form of acute nephritis marked by a special implication of the glomeruli there is an infiltration of leucocytes, and sometimes hæmorrhage in the Malpighian tufts.

By these changes the secreting structure, the tubules, their lining epithelium, and the Malpighian bodies lose in distinctness of parts (definiteness), and in the exactness of the proportions of their constituents (coherence). But by these changes they are clearly made more multiform, not more uniform, which is inconsistent with the formula of dissolution. This again instances that disordered heterogeneity which is often observed as one of the phenomena of dissolution. But the morbid matters introduced among the tissues have not, it will be noticed, the heterogeneity of elaboration and specialisation described by the formula of evolution.

In the organ as a whole there are like changes. The differentiated substance of cells is converted into albuminoid and fatty material, and the intertextural spaces and the protoplasm of matrix-cells are permeated by these degraded products and blood and serum.

In the third or atrophic stage the same order of metamorphosis results from the disappearance of tubules and glandular cells of epithelium.

The tissue-changes in the waxy and cirrhotic varieties of Bright's disease exhibit correspondences to the fatty transformation and atrophic stages of acute nephritis. But the amyloid degeneration of waxy kidney commences, not in tubules, but in the Malpighian tufts and arteriæ rectæ ; degeneration of the tubules is secondary to the vascular changes. The third stage of the amyloid form is one of atrophy.

Following Dr. George Johnson, F.R.S.,[1] we learn that in cirrhotic or contracted kidney 'the epithelium is opaque and granular, in a state of cloudy swelling ; the tubes are crowded and opaque with degenerated and disintegrated epithelium ; some tubes are deprived of their epithelium, some contracted, others dilated in various degrees, some lined by transparent uninucleated cells, others filled with organised fibrine, rarely with blood or with oil ; the basement membrane and the Malpighian capsules are thickened.' Malpighian bodies are completely destroyed in advanced cases, and represented by glistening spheres of connective tissue. To the naked eye the organ is contracted and its cortical substance diminished.

Quite evidently the essential lesion is a destruction of the cortical or secreting portion of the organs, and its reduction to relatively general, simple, and uniform fibrous tissue.

The Functional Changes.—Correlated with the rearrangements of structure are certain rearrangements of function, both in the kidneys and in organs that have physiological relations with them. To notice first the quality of the work done by the renal

[1] *Lectures on Bright's Disease.*

glands. This is made obvious in the urine. Within the range of healthy physiological action, there are present in the urine more or less precise proportionate quantities of water, nitrogenous matters, acids, salts, pigments, &c. In normal urine, too, there are such mutual relations among its constituents that the excretion when voided is a consistent body, being chemically stable under ordinary conditions, and, excepting a little mucus, holding its many ingredients in solution.

Can we say that the secretion of the kidneys met with in Bright's disease shows divergences from the normal corresponding in character with the associated structural divergences? The specialised substance urea, the highest product of tissue metabolism, is more or less diminished in percentage quantity in acute and cirrhotic nephritis, and probably in the later stages of the amyloid disease, where the secreting cells are only secondarily involved. Then there are present, with varying frequency in different cases, the substances albumens, globulins, peptones, extractives, urates, oxalates, &c., which are intermediate between urea and the food-materials—proteids, fats, and starches—from which urea is derived. Of these unspecialised intermediate bodies found in the urine in Bright's disease, serum-albumen is the most notable. Next there are casts of the uriniferous tubes, cells from the kidney and urinary passages, and commonly blood in the acute disease. And the urine of nephritis is relatively unstable, being often decomposed and malodorous at the time of evacuation. Lastly, there is the relative irregularity in quantity and

K

quality of the urine at different periods and under
different conditions of the organism. Now the excretion is abundant, and at another time scanty; at one
hour of the day there may be much albumen, at
another none; and similarly with the casts and
other abnormalities. Although acquiring a more
heterogeneous composition, urine so changed is
changed towards indefiniteness and incoherence, and
this stands for like changes in the renal functions.

Passing to the functional disturbances of parts in
physiological union with the kidneys, we may note
the inconstancy with which the bladder discharges its
contents. In health, vesical action is timed with
considerable exactitude to certain organismal conditions, as rest, work, and alimentation. But in
Bright's disease these relations are abrogated. Urination is uncertain, sometimes very infrequent, or for a
time in abeyance, in the acute disease; while in
chronic nephritis micturition is apparently simply
related to the presence in the bladder of an irritant
urine; those nice adjustments of the function to
social requirements and personal habits enjoyed in
health are thoroughly spoiled. The actions of the
bladder, in short, are less definite and co-ordinate.

The effusion of dropsical fluid into interstitial
spaces and serous sacs, which is met with most frequently in tubal or acute nephritis, may be considered
to obscure in some degree the distinct and orderly
heterogeneity of structures and functions; though
there may be considerable water-logging of organs
without obtrusive physiological perturbations.

Of other symptoms and complications of nephri-

tis that lend themselves to interpretation as dissolutional changes, anæmia is one. Here the blood
is despecialised and simplified by an increase of its
water and a decrease of its albumen and blood-corpuscles, especially of the highly differentiated red
corpuscles.

Gastric and intestinal catarrh, seen in all the
forms of nephritis, show us a diminished speciality,
complexity, and variety of functional performance.
There is digestive capacity for only the simplest and
most homogeneous of alimentary substances—a condition implying simplicity in the chemical and physical
processes of digestion ; and the gastric pains and dis-
comforts are the subjective concomitants of discords
among these processes. Normally, food, and the in-
digestible residue of food, pass in the definite direction from mouth to anus, and evacuations from the
lower bowel take place at regular times ; but in
vomiting, diarrhœa, and constipation, so often associated with nephritis, we see a subversion of this
definiteness of functional action.

Albuminuric retinitis, being histologically a fatty
degeneration and hæmorrhage in the retinal elements,
fully complies with the terms of the formula of dissolution. And by amblyopia, whether it depends upon
uræmia or retinitis, the functional attributes of the
organism certainly become less special, coherent, and
multiform. And as well by other nervous disorders
of Bright's disease—for example, loss of hearing, loss
of speech, or imperfections of speech, paresis, dyspnœa,
&c. Coma and convulsions, with their constituent
and accessory phenomena, loss of consciousness, dis-

ordered movements of limbs, Cheyne-Stokes respiration, &c., might easily be shown, were it required, to possess the characters of dissolution. It will not escape recognition that the pathognomonic high arterial tension conceals or erases the specific qualities of the pulse—the pulse ' wave,' peripheral outflow, and normal dicrotism.

Some of the inflammatory complications of nephritis, as pericarditis, bronchitis, pneumonia, and cirrhosis of the liver, will be noticed at length in later chapters.

But these are agreements only with the ultimate clause of the formula. Is there in Bright's disease a disintegration of matter and concomitant absorption of motion?

To begin with the first proposition, the disintegration of matter. This appears to describe the intrinsic nature of the transformations. Early in the course of the acute disorder, the stage of inflammation, there are found extravasations of blood and exudations into the interstitial spaces and tubules. That these involve disintegration was seen when we dealt with the phenomena of inflammation (chap. i. § 2). Tubules are crowded with cells of epithelium and fibrine, much of which becomes completely separated from the viscus, in the shape of casts. There is often a free proliferation of the nuclei of the epithelial cells— a phenomenon already expounded as virtually a breaking up of cell-substance. The urine usually contains blood, albumen, casts, leucocytes, and cells of epithelium; and these are portions of matter disunited from the body; they are the products of disintegrative actions. That disintegration continues through the fatty and atrophic stages is evidenced by the presence of cells and fat and casts in the urine.

In the lardaceous or amyloid kidney both molar and molecular disintegrations of the substance of the organ are unquestionably the intimate changes. Describing a typical example, Dr. George Johnson says: 'Most of the tubes were enlarged, and their epithelium was opaque from "cloudy swelling." . . . In some, the cells were in a state of fatty degeneration; and some tubes contained fibrinous coagula.' In other cases it is remarked that 'occasionally the tubular structure is rendered indistinct by an unorganised intertubular effusion. The Malpighian capillaries are thickened, opaque, glistening, and waxlike.' In the later stages the urine is often densely albuminous, and deposits casts and epithelial *débris*—disintegrated matter.

To be assured that the primary histological changes of cirrhotic nephritis are disruptive, the reader needs but reperuse what is said of this disease on a preceding page.

Turning to secondary and remote disintegrations of parts and functions correlated with the kidneys, the urine of nephritis is observed to be a less integrated and consistent fluid, insomuch that its constituents are often chemically dissociated, as proved by its premature decomposition. It has often an offensive odour when voided. Albuminuria implies a separation of the blood-serum, as also does the effusion of dropsical fluid. (See the definition of Disintegration, chap. i. § 2.)

Epistaxis, hæmatemesis, and other forms of hæmorrhage, are more obviously disintegrations; and catarrh occurring in Bright's disease, as gastric, intestinal, vesical, and bronchial catarrh, presents the disintegrations that inhere to the inflammatory process—exudation and cellular proliferation and exfoliation. And so do, in greater degree, the inflammatory complications pericarditis, bronchitis, retinitis, &c.

That the nervous phenomena of nephritis have for their physical bases a molecular or molar disintegration can hardly be doubted. Pain is the subjective side of physical disruptions;[1] and pathological anatomy testifies in numberless cases that paralysis and paresis are related in causation to destroying lesions in nerve-substance. It is probable that coma and convulsions in Bright's disease are due to chemical dissociations wrought by undepurated blood as it circulates in the higher cerebral centres, or by obstructed capillary circulation.

Disintegration of a kind differing from that just considered may finally be called to mind. The function of cleansing the blood of certain waste materials poured into it everywhere in the system is centred or integrated in the renal glands. Instead of each living cell, or assemblage of cells, having its own arrangements for excreting effete nitrogenous matters—an actual condition in the lowest organisms—the dis-

[1] See Mr. Grant Allen's *Physiological Æsthetics.* Henry S. King and Co. London, 1877.

tributing system—the vessels and their contents—compounds the entire community of cells, and so brings every part into functional union with organs specialised for the business of excretion. In some of the Protozoa, organisms without divisions of labour, there are found spaces in the protoplasm that are probably concerned in the excretion of water and other residual substances. The individual creatures in a sponge probably discharge their nitrogenous excreta into the water-canals which traverse the colony. In Platyhelminthes and Trematoda the renal function is distinctly organised, but is still diffused. We read of the excretory system of the latter, ' the finest ducts are distributed *through all parts of the organism*, and they pass into collecting vessels, which, by the formation of anastomoses, give rise to a most complicated plexus ; from these arise the efferent ducts, which gradually unite into collecting vessels.' Now in Bright's disease we find a sort of return to this primitive diffused form of the function in that other distantly connected organs probably assume vicariously the work of the impaired kidneys. The diarrhœa of chronic nephritis is generally attributed to intestinal efforts to discharge urinary substances. The serviceableness of purgatives in the treatment of Bright's disease bears out this view. A like meaning is attached to gastric catarrh and vomiting. And the presumption is a reasonable one that dropsical transudation is a still more general action, by which water and other components of the urine are stored up or cast out upon the intertextural spaces. In like manner, nasal and bronchial catarrh are probably depurating processes, and the inflammation of serous membranes may arise from the determination of unexcreted matters to surfaces where local conditions and constitutional peculiarities make excretion relatively easy. Thus the kidneys fail to do for the economy their appointed work, and the function becomes divided among the various co-operating parts of the system, and so becomes less integrated.

And now to learn whether this complicated series of disintegrations is accompanied by concomitant absorptions of motion. This calls out the question, What are the causes of Bright's disease? and our knowledge of these takes us at once to environmental influences.

All the various known conditions of the acute disorder include a poisoned state of the blood as the leading factor. In scarlatinal nephritis it is a necessary conclusion from the facts that excretion by the kidneys of vitiated blood induces the demolition of glandular substance. Doubtless enfeeblement of tissues, however occasioned, is often conjoined; but the substantial cause is the unhealthy matter in the blood seeking elimination by the kidneys.

Exposure to cold is said to stand next to scarlet-fever in point of frequency as a cause. Depression of the cutaneous temperature throws additional work upon the renal glands; but this can scarcely in itself be adequate to produce inflammation of them, and almost certainly co-existing depravity of the blood is the more efficient factor.

Blood rendered impure by imperfections of the

digestive processes, connected with excess or defect of alimentary material, is a very common cause.[1] Then there is the acute nephritis of diphtheria, measles, acute rheumatism, erysipelas, pyæmia, malaria, typhoid and typhus fever ; and of the parturient state, in which the blood to be cleansed is unduly charged with excremental matter derived from the fœtus or absorbed from the uterus. And clearer instances are obtained in poisoning by carbolic acid, cantharides, and turpentine. These causes of acute Bright's disease show in every example that the disintegration of matter in the kidneys is the consequence of energy absorbed from the environment, however complex and extended the chemical and physical events intermediate between the incidence of the external force and the production of the special internal lesion.

We do not yet know all the causes of chronic nephritis, but those now recognised are co-natural with the causes of acute nephritis. The large white kidney sequential to the desquamative nephritis of scarlet-fever or other zymotic disease, with great likelihood follows from the inability of the glands, weakened and disorganised by previous inflammation, to do the work entailed by the organism's resumption of normal dietetic and other habits. According to Dr. George Johnson, large white kidney may be related to excessive eating and drinking. He has known it as the sequel of tropical malarious fever, and of ague, which are blood diseases beyond question.

The ætiology of lardaceous kidney is obscurely related to constitutional syphilis, protracted suppura-

[1] Cf. Dr. George Johnson, *Op. cit.* p. 40.

tion, and other states of the system that may be referred to outer actions of one kind or another.

Common causes of cirrhosis of the kidneys are alcoholic intemperance and intemperance in the consumption of food. It is probable that excessive eating is as frequently the cause of hepatic and renal disease as excessive drinking'[1]—an opinion my own experience obliges me to endorse. Less frequently, gout, diabetes, and lead-poisoning produce the cirrhotic disease. Gout and diabetes, as we may possibly see subsequently, are causally connected with the alimentary environment; and the nephritis co-existing with these diseases is generally regarded as dependent directly upon the elimination in the one case of uric acid and allied substances, and in the other of sugar.

I think the available data regarding the ætiology of nephritis support the conclusion that with the disintegration of the structural elements of the kidney there is a concomitant reception of energy from without.

Of the secondary changes in other organs and tissues, the intimate and immediate cause of the disintegration can be safely assumed in only a few instances. The inflammations of serous and other membranes may be supposed to be directly connected with absorptions of chemical energy from the blood. Hæmorrhages in Bright's disease result from increased intravascular pressure and degeneration of vascular tissue, and these are effects of abnormal states of the blood, as, in all probability, are many of the nervous

---

[1] *Urinary and Renal Disorders,* by Lionel S. Beale, M.D., F.R.S.

phenomena of Bright's disease. Apoplexy and paralysis are secondary to cerebral hæmorrhage.

The Evolutional Changes.—Acute nephritis terminates in complete recovery in a large number of cases. With the removal from the body of the causes of its disturbances the dissolutional processes are succeeded by those of resolution and repair (chap. ii.). Our knowledge of the rearrangements of matter and motion leading to recovery is almost wholly inferential. Irritant substances are probably slowly eliminated by the excretory channels, or are temporarily disposed of with the fluid of effusions in the interstitial spaces, these adaptive changes being often furthered by the art of the physician.

There are no direct observations to prove that the destroyed portions of the kidney are regenerated, but we may rest upon the inference that such is the case. Since the kidneys ultimately resume their normal work, the matter—cells and serum—which was exuded through vessel-walls into the connective-tissue spaces and tubules must be either cast out of the body, which would complete the disintegrative process, or it must be resorbed, which resorption we saw to be an integration of matter and dissipation of motion. Renewal of the epithelial elements of the glands is also an evolutional process.

One of the earliest signs of returning health in acute desquamative nephritis is the disappearance of dropsical effusions. Here, again, the effused matter reaggregates (integrates) within the vessels, restoring (dissipating) that motion which was imparted to it at its effusion. And to replace the losses by disintegration there is ingested fresh integrable matter in the shape of food. Seeing that recovery means the organism's return to a normal state, functions and structures regain their lost speciality, organisation, and multiformity.

Chronic nephritis is held to be incurable ; consequently we cannot speak of resolutional operations. But there are well-recognised reparative and adaptive changes. As cardiac hypertrophy and arterio-capillary fibrosis are conjoined in causation, they may be referred to together. A careful appraisement of all the facts that bear upon the difficult question as to the conditions out of which they arise almost convinces me that enlargement of the left ventricle and thickening of coats of the systemic arteries are joint effects of a single cause. This is an alteration in the quality of the blood. It usually depends, I think, on slight failure of excretory power in the kidneys long anteceding distinct signs of Bright's disease; or it has a different origin. But on account of it the blood passes with difficulty through the finest conduits of the arteries, and thus there is imposed an unnatural resistance to the onward flow of arterial blood. To meet the obstruction, the coats of the arteries and the left ventricle become hypertrophied, and this induces high arterial tension. These we must include among the minor evolutions in the general dissolution of Bright's disease (chap. iii. § 1).

Others are seen in the new growths of connective tissue in the kidneys themselves. In acute nephritis, when somewhat prolonged, as well as in the chronic disease, there is an accession of fibroid material. The walls of the small arteries, the capillaries, the Malpighian tufts, Bowman's capsule, and the basis sub-

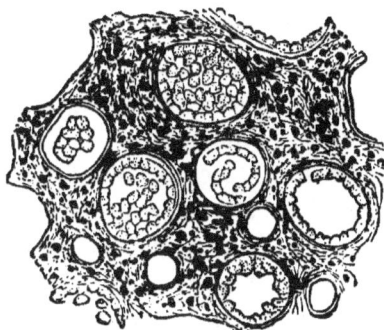

FIG. 4.—TUBAL NEPHRITIS.

Duration of disease, six months. Kidneys large; capsules, non-adherent; surface, smooth; tissue, soft. Showing, in addition to the intratubular change, the cellular infiltration of the intratubular connective tissue.   × 200.

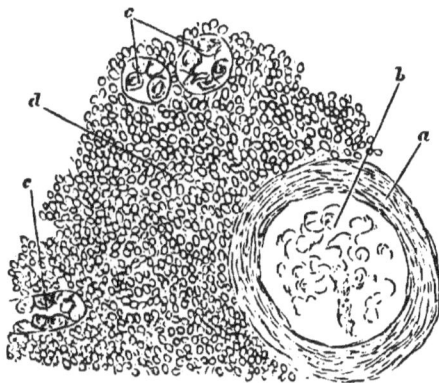

FIG. 5.—INTERSTITIAL NEPHRITIS. SCLEROSIS OF GLOMERULUS.

a, thickened capsule; b, condensed vessels of tuft; c, dwarfed tubules remaining; d, abundant round cells in interstitial tissue.   × 350.

stance of the tubules, are seen to have grown. The annexed drawings (figs. 4 and 5) are, in part, demonstrations. According to the view we have submitted (chap. i. § 1, pt. ii.), this increase in the connective tissue is explained as an effect of loss of parenchyma, and not of inflammation from irritant blood acting directly upon the interstitial substance. It was seen a moment ago that the destination of the irritant blood is the proper excreting substance. We must, then, number it among the evolutional changes of Bright's disease. Observation of the figures and specimens, guided by the view proposed, will satisfy the reader that this added substance has the symmetrical and orderly characters of a reparative structure, that it advances always in the direction of diminishing resistance to growth caused by removal or death of the parenchyma, and that it is never unaccompanied by disease of the parenchyma. Senile renal atrophy, the later periods of acute nephritis, the later stages of amyloid disease, and chronic cirrhosis, one and all bear testimony. Why this parenchymatous change is never absent we shall now be able to understand more completely.

The Lesions as related to the Developmental Rank of the Tissue-Elements.—Is the order in which the histological parts are implicated in the dissolutional process unisonous with the tenet that the most specialised parts are the least stable ? (chap. i. cor. 1, pt. ii.). Mark first the broad fact that in all forms of hæmatogenous nephritis the pathological changes begin in those elements deputed to do the special work of the organ. Though contrary to the usual explication of the changes, in renal cirrhosis the disease *begins* in the secerning portion of the glands—the cortex. This is the opinion of Dr. George Johnson[1] —an opinion borne out, I think, by every example. (See figures and text on preceding page.)

Mark next that the separate elements of the secerning portion are involved according to their degree of speciality. With the epithelial structures it is first the epithelial cells, then the tubules, and the more specialised convoluted tubules before the straight tubules, that are affected. With the vascular structures, it is the very specialised expansions of the nephritic vessels, the Malpighian tufts, and smallest capillaries that prove most unstable. This gives the distinguishing morphological features to glomerulo-nephritis and amyloid kidney; but acute tubal nephritis and granular kidney do not supply less perfect exemplifications.

Interpreted in this way, the several kinds of hæmatogenous nephritis, in which the incident de-structive force is somewhat equally distributed, arise in chief part from differences in the intensity, duration, and nature of the irritation caused by the depraved blood. In each kind it is the most differentiated portion of the gland that undergoes dissolution. If the irritation is severe and transitory, we may have the phenomena of acute nephritis, with damage to specific elements and a subsequent restoration of functions and structures; if the irritation is of mode-rate intensity and long-continued, there is a gradual effacement of the specific elements, and a reproduction of connective tissue to replace them. In such way arises the atrophic kidney after acute nephritis, after

---

[1] *Lectures on Bright's Disease.*

amyloid nephritis, and after chronic desquamative nephritis. The higher and accessory structures and functions pass away, but the lower and fundamental continue to live. And thus the views that parenchymatous disease is primary, and not secondary, to interstitial growth, and that the highest elements have the least stability, are views sustained by the facts, and are mutually sustaining.

We must not linger over other affections of the kidneys and urinary system, useful as it might be to apply the principles to them. It may be incidentally pointed out that calculus disease presents phenomena paralleling the phenomena of blood-coagulation (chap. iii. § 1). The aggregation of the matter of a calculus is an evolution intercurrent with a larger dissolution. Perfectly healthy human urine holds its varied constituents in a state of solution, and the precipitation of these within the body may be regarded as a decomposition of the urine, the formation of a calculus being a redecomposition. In the common cases of stone this unbalancing of the chemical composition of the water can generally be shown to proceed from physiological disturbances that possess the generic characters of dissolution. And there are facts at hand arguing the causal connection of these disturbances with the inception of surrounding energies. But many links in the chain of causations remain to be discovered.

§ 3. CIRRHOSIS OF THE LIVER.

The Structural Changes. — Chronic interstitial hepatitis serves most conveniently as a test of the principles we are employing. The changes of morphology and physiology are well-defined ; the disease is a familiar one, and its ordinary cause is simple, and known with certainty.

To see that the general transformations of structure are in keeping with dissolution, we may first recall the normal histology of the gland. Briefly, hepatic tissue consists of an assemblage of specialised epithelial cells separated more or less distinctly into lobules or acini by connective substance. This intersecting connective substance forms a kind of trellis for the support of branches of the hepatic artery, hepatic duct, portal vein, lymphatics, and nerves.

Looking at the results of well-marked common multilobular indurative hepatitis, we perceive the following alterations. The hepatic cells with their albuminous cement and the intercellular vessels are destroyed in great numbers ; whole lobules are blotted out, and others remain as mere collections of fat. One of the intervening changes appears to be the conversion of hepatic cells into oily matter. Coincidently there is an overgrowth of connective substance. In addition, bile-ducts and portal veins are narrowed and obstructed or annihilated.

Throughout the tracts of interstitial tissue, and often at the margins of the lobules, lie numbers of inflammatory corpuscles, and drops of oil are interspersed among the glandular epithelia. And the

L

nuclei of the latter are obscured owing to their degeneration.

Thus, the architectural features become less recognisable, less definite ; the gland becomes disorganised, and the substitution of connective tissue for hepatic cells, bile-ducts, and vessels clearly constitutes a change to relative simplicity and generality.

The Functional Changes. — The physiological counterparts of these lesions are found among the symptoms of cirrhosis. Of course many of the symptoms are caused by the direct influence upon the stomach and other organs of the forces that produce hepatitis. Including any of the chief disturbances of function, these are seen to bear the stamp of dissolution. The multitudinous, orderly connected, and decided functions of the stomach are either suspended or performed very indeterminately. The contractions of the stomach, in health definite in time and direction as related to the taking in of food, are so perverted that food returns through the mouth, or the contractions take place as in ' morning sickness,' independently of the normal excitants to contraction.

Diarrhœa evidences the same order of changes in the intestinal functions; and ascites and œdema show how the infinitely nice and varied circulatory actions going on in the normal liver are greatly disordered or have ceased altogether. Jaundice tells of identical mutations in one of the gland's special functions. But these correspondences with dissolution, which might be readily extended, are less interesting than the disintegrations of matter and concomitant absorptions of motion.

Disintegration of matter is seen in the degeneration and fatty metamorphosis of liver-cells and intralobular vessels, and in the subsequent atrophy of them. Secondary disintegrations are found in the effusions of dropsical fluid, in diarrhœa, hæmatemesis, and other accompaniments of cirrhosis. And along the entire series of disintegrations the concomitant absorptions of energy can be followed out.

Supposing the cause of the interstitial hepatitis to be the common one, an intemperate consumption of alcohol, then, beginning with this chemical energy we infer that it is first absorbed by the veins of the stomach. As it enters next the substance of the liver by the channels of the portal system it is conducted along the connective tissue to the hepatic lobules. As we have chosen to decipher cirrhotic changes ·(chap. i. §§ 1 and 2, pt. ii.), the alcohol setting up chemical dissociations destroys the proper liver substance, and the interstitial tissue grows out to fill its place. Thence backward upon the portal and abdominal veins, as the portal capillaries and ducts become defunct, motion in the form of intravascular mechanical pressure gives rise to the effusions of ascites, or diarrhœa, or hæmatemesis. Or bile is forced into the blood, producing jaundice. And so on with the remaining disintegrations : they are all, including those attendant upon inflammation (chap. i. § 3), caused by the incidence of energy referrible more or less remotely to the energy ingested as alcohol.

The Evolutional Changes.—There are few well-authenticated cases of recovery from unmistakable cirrhosis. It is conceivable, and indeed to be expected, that in the early stages, as the action of the cause ceases, and other conditions become favourable, resolution, or even the reproduction of true liver-tissue, might occur.

A redevelopment of bile-ducts is spoken of by French observers ; but others suppose the ducts to be dilatations of pre-existing bile capillaries. Adaptive developmental changes are found in new venous communications between the portal and general systems. The vein of Sappey, connecting a branch of the portal and the epigastric and internal mammary veins, is an example. And there is said to be occasionally a temporary compensatory growth of hepatic arteries to meet the cutting off of portal blood by the destruction of portal capillaries.

Lastly, the new formation of interstitial connective tissue may be again cited as a reparative process. We see that in hepatitis as in nephritis the noxa is carried to and most freely dispersed among the glandular cells, not to the ramifications of the capsule of Glisson, whose blood-supply is from the hepatic arteries. The proper substance is the destination of the portal blood as far as the liver is concerned with it. And in many cases the anatomical changes make it demonstrable that the disease begins in the parenchyma. Specimens of cirrhosis are met with showing parenchymatous degeneration without interstitial growth. There are, on the other hand, some forms of hepatic disease that apparently militate against this rendering of the changes. The atrophy of glandular cells in simple and pigmentary atrophy is not attended with marked increase of interstitial substance. Here, however, as in certain cases of cirrhosis, it is highly probable that nutrition is too low to admit of connective-tissue growth. Simple hepatic atrophy unattended with augmentation of interstitial substance is one of the consequences of starvation.

If so-called 'hypertrophic cirrhosis' presents as stated [1] an interstitial hyperplasia, without equivalent atrophy of lobules, this would conflict more seriously. But not improbably the hepatic enlargement is due to inflammatory exudation and hyperæmia, and this variety may really be an early stage of ordinary cirrhosis.

The Lesions as related to the Developmental Rank of the Tissue-Elements.—Like Bright's disease, ordinary indurative hepatitis is hæmatogenous, and therefore the tissue-elements are under nearly equal strains from the action of the destroying force. Can we say that the highest structures are the least stable ? We must first think of the liver as accessory to the intestine. Its embryological history teaches that the gland arises as a differentiation of the enteron ; in vertebrata budding out from the duodenum upon the abdominal cavity. Thus it is an appendage to the alimentary canal, and this may in part explain its relative susceptibility, as compared with the stomach, to the disintegrating influences of alcohol. When cirrhosis is caused by intemperance the stomach must sustain the stress of the irritant ; yet gastric fibrosis does not appear to have been observed as often associated with hepatic fibrosis (cirrhosis). That the liver rather than the stomach is destroyed must be determined by several circumstances ; but it seems likely that one of these is the *greater stability of the*

---

[1] Ziegler's *Text-book of Pathology.*

more fundamental gastric tissues. Whether the loss
of acini is occasioned, according to the explanation in
vogue, by atrophy secondary to growth of connective
tissue, or, as I prefer to think, by the direct action
of alcohol, the fact remains that the specialised portions of the gland are alone destroyed. Note the
order of their destruction. The changes begin at
the peripheries of the lobules. Judging from cases
in the early stage, the nuclei of the epithelial cells
disappear sooner than the protoplasm. The cells lose
their distinctive polygonal shapes and become simple
spheres ; then the protoplasm loses its specialised
chemical qualities, as evidenced by the action upon it of
staining-agents. Bile-ducts and vessels are more persistent, but in the end common connective tissue and
simple fat are almost the only surviving elements.

If we remember the symptoms and morbid anatomy
of biliary cirrhosis, simple hepatic atrophy, acute yellow
atrophy, passive hyperæmia of the liver, and other
diseases of this organ, no doubt will be felt as to the
possibility of making an equally satisfactory synthesis
of the facts. It will receive attention that in acute
yellow atrophy from phosphorus-poisoning, the destroyed elements are the glandular cells, though the
poison may greatly injure the connective substance,
which is unable to withstand its intense action. The
energy is absorbed by radicals of the portal veins. In
passive hyperæmia of the liver (nutmeg liver) there
is again partial extinction of the proper liver substance, the energy—pressure of venous blood—being
transmitted by way of the hepatic veins.

§ 4. INFLAMMATION OF THE LUNGS.

Under this heading may be noticed the most important affections of the lungs. These are croupous, catarrhal, and interstitial pneumonia, and phthisis.

The Structural Changes.—Important as are the clinical, and certain as are the pathological, differences in these forms of pulmonary inflammation, they cannot radically be separated from one another. The dissimilarities in the textural changes are dissimilarities in details merely.

In the case of each the minute specialities of structure in the tissue of the lung are decreased, confused, and rendered less distinguishable. And these transformations are effected in quite corresponding ways. We may notice first the morbid anatomy of croupous pneumonia. This begins as inflammation ; the vessels of the parenchyma dilate, and serum and cells pass out of them (chap. i. § 2). This constitutes the stage of engorgement. In the next stage, that of red-hepatisation, the extravasated matter has filled the pulmonary alveoli, still further cancelling the structural differentiations. Additions to the cellular elements of the exudate and its degeneration and disintegration mark the stage of grey hepatisation. If the process terminates in abscess or gangrene, the changes of dissolution are then very pronounced.

Broncho- or catarrhal pneumonia is also an inflammation of the parenchyma with inflammation of

the bronchi superadded ; and it differs histologically
from croupous pneumonia only in the qualities of the
exudate and in the concomitance of collapse and
emphysema.

' Interstitial or chronic pneumonia is charac-
terised by a gradual increase in the connective tissue
of the lung, which leads to an induration of the
pulmonary texture, and to progressive obliteration of
the pulmonary cavities.' As 'in interstitial nephritis
and hepatitis, by this destruction of pulmonary parenchyma and growth of fibroid material the lungs are
disorganised and reduced to relative uniformity and
generality.

Phthisis derives its distinctive characters from a
particular combination of the foregoing changes. It
is an inflammation of the parenchyma with accumu-
lation of fibrine, leucocytes, and epithelial cells in
the alveoli ; there is conjoined a thickening and
infiltration of the walls of the alveoli and terminal
bronchioles, and a growth of the interlobular connective tissue. These changes, broadly considered,
are a criterion of dissolution ; and so in a greater
degree are the changes of coagulation-necrosis,
softening, ulceration, and excavation, which belong
especially to phthisis. The consolidation of tissue
associated with inflammations of the lungs, when
simply the consequence of effusion into the alveolar
spaces, is not strictly an evolutional change or integration of matter ; as just shown, the tissue is
really disintegrated. But the coagulation of the
effused substances, their occasional organisation, and
the concretion of interstitial matter, are secondary

evolutions intercalated with the primary dissolution.

The Functional Changes.—Reserving for later consideration those dynamic phenomena which inflammation of the lungs shares with the fevers, it will be sufficient to note of the functional correspondences to dissolution that with dyspnœa, irregular and hurried breathing, and cough, the respirations of health lose their determinate and orderly qualities. And the pulse-respiration ratio being disturbed, the respiratory rhythms cease to be less consentaneous with the cardiac rhythms. These perversions of organic rhythms correspond with perversions of function in the pulmonary tissue correlated with the observed structural perversions.

It is evident, both from the lungs as seen microscopically, and from the clinical phenomena, that these diseases are essentially a disintegration of the body. Cells and serum are separated from the blood, and the alveolar epithelium desquamates. There are the disintegrations of the fevered state ; and in croupous and catarrhal pneumonia rusty and bronchitic sputa. Phthisis is an obvious ' wasting away ' ; the waste matter being thrown off abundantly as sputum, as hæmorrhages from the lungs, as diarrhœal discharges, as sweats, &c.

But is it possible to connect these disintegrations with absorptions of energy as their causes ? If we could accept the assumption that micro-organisms— the *pneumococcus* and *bacillus tuberculosis*—are the intrinsic factors this would be possible and easy.

But, the assumption being held as not consonant with
the varied clinical and pathological facts, can a different ætiology be suggested ?

In the special diseases discussed in the immediately preceding sections the data clearly prove that
the parenchyma of the kidneys and liver is destroyed
by matters received from the blood. Bright's disease and cirrhosis of the liver are said to be hæmatogenous. Now in the case of the kidneys we see
that there are two chief paths by which noxæ are
conducted to the renal substance. One of these is
along the renal arteries, and the other is by extension
of morbid processes from contiguous parts, as when
the renal disease is metastatic or caused by the mi-
grations of micro-organisms. In the case of the
liver, if we exclude obstructions to the outflow of
bile, the ways by which noxæ arrive are three—
namely, by the portal vessels, by the hepatic arteries,
and by metastasis. But with the lungs the routes
are four. Irritants may come with the air breathed,
by way of the pulmonary arteries, by way of the
bronchial arteries, and by continuity of tissue. That
the irritants producing ordinary pneumonia and
phthisis do not take the two last-mentioned courses,
may be concluded without misgivings. There remain,
then, the air and the blood conveyed by the pulmonary artery. Inhaled impurities are beyond all doubt
the causes of many lung affections. The mention
of stonemason's cirrhosis, of grinder's rot, and of
miner's lung will recall many more kindred examples.
Or the inhaled impurities may be gaseous or organic.
Breathing nitrous fumes will cause extreme irritation

of the lungs ; and purulent and other secretions of the mucous membranes of the air-passages are often inspired, producing infective inflammation and collapse of the pulmonary parenchyma. It is almost certain that insufflated bronchial secretion containing tubercle bacilli has a vicious influence in the late stages of phthisis. The relations of the lesions to the branches of the bronchi are indicative of this. Bronchopneumonia may be caused by inflammatory products being drawn into the air-vesicles. And the *micrococcus pneumonia* and *bacillus tuberculosis* may reach the pulmonary tissue with the respired air. But these contributory and extraordinary sources of pneumonia and phthisis have not been accepted by physicians and pathologists as sufficing for the daily occurring forms of these diseases. The fact that tubercle bacilli are inhaled in the hospitals for consumptives by persons who do not contract phthisis shows that the micro-organism alone is inefficient. And states of the atmosphere, as east-winds, are not active with all persons in assisting to produce pneumonia. Efficient or co-efficient causes must, then, have some other source. What, probably, is this ? I think the blood distributed to the pulmonary parenchyma by the pulmonary arteries. It has not been dwelt upon by writers that the blood coursing through the alveolar and bronchiolar ramifications of the pulmonary artery bears to the parenchyma of the lungs relations analogous to those which the blood of the renal arteries and the blood of the portal veins bear to the parenchymas of the kidney and liver. The lungs may be said to exist for the blood of the

pulmonary arteries as the kidneys and liver exist
for the blood of the renal arteries and portal vein.
Should we not, then, look to the pulmonary arterial
blood as the source of the chief disturbances of function and structure in the pulmonary parenchyma ?
The proper substance of the lungs is exposed on
the side of the pulmonary arterial blood to incident forces far surpassing in magnitude and variety
the forces incident from other sides.   And this truth
is accentuated by the consideration that the lungs
have almost immediate relation with every ounce of
digested food.   After elaboration by the processes of
digestion alimentary matter is then passed through
the portal vessels to the parenchyma of the liver, and
through the lacteals to the receptaculum chyli.   It
is very probable that the bulk of the food-material
enters the portal circulation ; but the point to be em-
phasised is, that after leaving the liver and receptaculum nutrient substances are at once carried to the
lungs.   Before being subjected to further chemical
and physical changes they are brought into the
closest converse with the pulmonary parenchyma
through the blood of the pulmonary artery.   The
parenchyma of the lungs, indeed, is scarcely more
than a plexus of pulmonary capillaries.   How profoundly and constantly, then, must the proper tissue
of the lungs be ordered by the amounts and kinds of
the food ingested !   As profoundly and constantly,
probably, as the life of the proper tissues of the liver
and kidney.

For these and other indirect reasons, and directly
from my own observations in the treatment of pul-

monary affections, it appears to me that ordinary
pneumonias and phthisis are most intimately connected in causation with the quantity and quality of
the pulmonary arterial blood *as related* with the
quantity and quality of the air respired.   But of
these two sets of related and reciprocal factors, I
think the former is at least of equal practical importance with the latter.   Giving as concrete an expression of the view as will be proper in the absence of a
full exposition, it may be said that both pneumonia
and phthisis are initially dependent on the action of
pulmonary arterial blood made irritant by the ingestion of superabundant, improper, or impure food ;
closely associated conditions being a relatively feeble
respiration of irritant air.

It is certain, however, that if the principal factors
are thus circumscribed there are in every actual case
many accessory determining and predisposing factors of much moment, and that out of their interrelations arise the differences among cases and the
special features of each disease.   The infectiveness of
phthisis, its clinical history, certain features of its
pulmonary lesions, as their lobulated character and
apical distribution, imply a distinct unlikeness of
causative conditions to those obtaining in pneumonia.
But, as here suggested, an important likeness underlies
this unlikeness.   At what stage of phthisis the specific organism becomes active, whether it is usually
received by the food or by the air, it is at present
impossible to say.

Would it not introduce a too lengthy digression,
it could be shown that the ætiology of pneumonia

and phthisis so conceived may be the means of
rendering more intelligible several classes of facts.
It is a matter of common observation that phthisis
frequently begins as a disease of the digestive organs,
and is accompanied throughout its course by deep
disturbances in the functions of these organs.   The
dietetic and other habits of those suffering from
phthisis, and the commencement of the disease at the
apices of the lungs, would acquire a better meaning.
And so would the clinical phenomena and earliest
histories of cases of pneumonia, the signs of plethora
and indigestion, and the precursory nasal and bronchial catarrh.   And the pneumon·as secondary to
blood-poisoning and other conditions could by this
ætiology be assigned a natural place.

Granting for the nonce that this is a general presentation of the truth, the blood of the pulmonary
artery acts as a phlogogenic agent, and irritants being
present in the air or in the secretions of the respiratory tract, the absorbed energy sets up the various
disintegrations.   The alveolar membrane is denuded
of its epithelial cells ; serum and leucocytes pass out
of the vessels into the alveolar and trabecular spaces,
and, the process becoming chronic—the causes per-.
sisting—there results such destruction of bronchioles
and alveoli as is observed in fibrosis and phthisis.
And so with the less common forms of pulmonary
inflammation, pneumonia by metastasis, pleurogenous
pneumonia, and pneumonia caused by the inspiration
of mechanical and other irritants ; the disintegrations of matter can be traced to concomitant absorptions of energy.

The Evolutional Changes. — Resolution is the usual ending of both croupous and catarrhal pneumonia ; and the absorption of intra-alveolar exudations and the recovery of portions of pulmonary tissue is a process recognised as often occurring in phthisis. And the process of evolution in the alveoli is repeated on a larger scale by the systemic processes of recovery ; the functions of respiration, circulation, alimentation, &c., become less and less inexact and incoherent until the normal state is re-established.

The explanation we have proposed of the hyperplasia of connective tissue in cirrhosis of the liver and kidneys will be found applicable to the indurative processes of pneumonia and phthisis. These processes are then evolutional. It may here be remarked, as showing how fibrosis follows from death of the parenchyma, that in atelectasis and collapse, which lead to simple suppression of function and structural atrophy of the proper pulmonary substance, there is but little attendant inflammation, yet often considerable resulting induration. As in similar diseases of other organs, nutrition greatly influences the growth of connective tissue in the lungs.

The Lesions as related to the Developmental Rank of the Tissue-Elements.—Observe that notwithstanding the differences in their ætiological conditions and morbid anatomy, pneumonias and phthisis are dissolutions of the specialised parts of the lungs. And when there is reason to think that the destroying forces fall equally upon each class of elements, the process begins in the highest elements. This is plainly the teaching of the morbid anatomy and physiology and of the facts of ætiology. In croupous, catarrhal, and interstitial pneumonia it is the alveoli that are first affected, and in phthisis the alveoli and bronchioles. The terminal expansions of the pulmonary arteries, and the epithelium of the alveoli, alveolar ducts, and respiratory bronchioles, are the parts that suffer disintegration. By secondary processes of phthisical infection the tissues may, of course, be attacked in different order. In advanced cirrhosis the smaller vessels and bronchioles are obliterated ; and in advanced phthisis the fundamental tissues, bronchial and connective, may secondarily be excavated and destroyed.

The successive implication of the higher and lower structures is indicated by the tracts of fibroid tissue in cirrhosis and phthisis. First the alveolar, then the interlobular, and, lastly, the peribronchial connective tissues are thickened.

The remaining pneumonic morbid processes will not, I think, be found on examination to offer insuperable obstacles to interpretation by these principles. Brown induration of the lungs shows us disintegrations of the higher pulmonary elements from the transmission of mechanical pressure through the blood of the pulmonary veins ; also, hyperplasia and integration of connective substance as a sequel of these lesions. Broncho pneumonia from inspired irritants closely repeats the phenomena of pulmonary cirrhosis. Emphysema subscribes to the principle of dissolution in that the least-organised capillaries and alveoli disappear from innutrition, and mechanical pressure transmitted through the air. And when the morbific forces are distributed mainly about the bronchi, the ensuing bronchitis is marked by damage to special structures and hyperplasia of connective tissue.

CHAPTER II.

SPECIAL DISEASES AS EXEMPLIFYING DISSOLUTION
AND EVOLUTION CONTINUED.

§ 1. LOCOMOTOR ATAXIA AND OTHER DISORDERS OF THE NERVOUS SYSTEM.

The Structural Changes.—That we may appreciate the traits of dissolution in sclerosis of the posterior columns of the spinal cord, we must note the changes prior to loss or atrophy of its substance. If, as is believed, the atrophy commences in the posterior root-zones, or columns of Burdach, then the nerve-fibres of this group, before they disappear, exhibit the metamorphosis of degeneration (see chap. iii. § 5). The medullary sheaths and axis-cylinders are broken up, and fat granules, corpora amylacea, and nuclei appearing, the natural order, distinctness, and unlikeness of the histological components give place to an unarranged blending of simple elements. And when, later, the parenchymatous tissue has passed away and only the overgrown neuroglia remains, the change is visibly from the heterogeneous to the homogeneous, from the complex to the simple, and from the special to the general. Also, the alterations of structure are identical if the degeneration involves other systems of nerve-fibres or nerve-cells, such as the columns of Goll, the posterior roots, or the posterior and marginal portions of the lateral columns and the posterior horns of grey matter.

The Functional Changes. — The physiological manifestations in nervous disease evince dissolution in terms very readily understood. Of the sensory phenomena of tabes dorsalis, the various paræsthesiæ—girdle pains, the feeling as if the soles of the feet were resting upon velvet, hypersensitiveness to temperature and insensitiveness to touch, analgesia or polyæsthesia, and loss of the muscular sense, mean that sense-perceptions have lost in precision, that impressions reach the sensorium less bound together (incoherently), and that sensations have become relatively homogeneous. The motor phenomena are of the same order. Inability to maintain the erect posture with the eyes closed or open is known to be due, as the term static ataxia denotes, to an imperfect co-ordination of outgoing or ingoing currents, either in the centres of the spinal cord or at the seat of consciousness. And this ataxia or incoherence is shown in walking, in prehension, in speech, and in the movements of the eyes (strabismus nystagmus), and is always accompanied by its token, indefiniteness of movement. When paresis and paralysis of the limbs or bladder, amaurosis, loss of sexual power, &c., supervene, functions are then most obviously less heterogeneous.

The microscope has shown us that disintegration is an inalienable character of the spinal lesions. Can this disintegration be connected with the imbibition of energy ? It is not yet settled what conjunction of conditions is productive of locomotor ataxia. Syphilis is common but not invariable in the histories of these cases. But the morphology of the spinal changes, and their restriction to functionally related groups of neural elements, do not harmonise with the supposition that syphilis is the supreme causative condition. The supposition may with little risk be rejected. It will be useful in endeavouring to arrive at the causes of tabes to consider the principal avenues by which disintegrative forces reach the elements of the spinal cord. We may distinguish three. One is through anatomically related parts by direct extension ; another is by the blood-vessels ; and a third by the spinal and cephalic tracts of neural tissue which, through the sympathetic and peripheral nerves, bring the organism into converse with its environment. That tabes does not arise from the extension to the spinal cord of pathological changes in contiguous tissues, we need not stop to demonstrate. Is the disease, then, hæmatogenous ? Analogy makes it conceivable and possible that noxæ distributed to the spinal cord with the blood might by a selective chemical affinity pick out and destroy limited districts of nerve matter. But, excluding syphilis, there are no evidences of specific blood derangement in the clinical records of typical posterior sclerosis ; and it appears unlikely that the disorder is inflammatory. Therefore the ascription of tabes to irritant blood may be discountenanced. Is the disease, then, attributable to influences reaching the spinal cord through the nerves ? Such appears from the facts to be actually the case. If locomotor ataxia is regarded as a simple degeneration from excessive functional action of distinct tracts of neural substance, the morbid anatomy, symptoms, and clinical histories of cases of the disease are not irreconcilable. As in all equally complex diseases, various secondary conditions must be placed among the factors ; but the chief factor of ordinary locomotor ataxia, many facts unite to show, is undue functional activity of the sexual system. Admitting provisionally that the balance of probabilities is at present in favour of this origin, then the nervous stimuli, having their source in sexual excitement and other nervous and psychical actions, impinge upon definite areas of the spinal marrow, and the energy absorbed in excess works the characteristic disintegrative changes. Destroying the posterior root-zones, there is prevented the transmission of duly proportioned afferent impulses to the cerebellum, with resulting ataxia ; impinging upon the posterior nerve-roots, there is caused first irritation of these and lightening pains and subsequently tissue destruction and anæsthesia. The posterior grey horns being injuriously affected, the various paræsthesiæ ensue, and muscular atrophy is thought to come from degeneration of the ganglion cells of the anterior horns. Paralysis follows disintegration of the pyramidal tract, and so on with the remaining lesions and symptoms. It is extremely probable that the disintegrations of ordinary locomotor ataxia are caused by absorbed nervous energy having correlations with the energies of the outer world.

The Evolutional Changes.—Recovery from the less

severe forms of tabes is frequent. If function is restored after the onset of structural derangements, we
must suppose either a reproduction of nerve elements,
or the delegation of the lost functions to surviving
nerve systems. Muscular adjustments, to maintain
equilibrium in walking and other movements, must
be counted among the adaptations. The evolutional
change of most interest to us is the hyperplasia of
neuroglia to repair the loss of nerve matter. Happily,
in holding that degeneration of the spinal parenchyma is the primary change, we are not wholly in
antagonism with the opinions of pathologists. Ziegler
remarks : ' Some have regarded the neurogliar hyperplasia as the primary disorder, and the degeneration
of nerve substance as secondary to it. There is, however, no real doubt that the degeneration is the primary and essential feature of the disease.' That it is
so is entirely confirmed by the ætiological considerations here presented, if these contain the truth, and is
upheld by the considerations which follow.

*The Changes as related to the Developmental Rank
of the Tissue-Elements.*—If the disease is properly a
destruction of the specific elements of the cord, with
overgrowth of neuroglia from the taking away of
normal impediments to its growth, posterior spinal
sclerosis amalgamates with sclerosis of the kidney,
liver, and lungs. The higher elements perish and
the lower endure.

And the details of the retrograde process are
fulfilments of the corollary from dissolution, as
Dr. James Ross has pointed out in his ' Diseases of

the Nervous System.' Ganglion cells first lose their elaborate processes, changing from bodies with caudate prolongations to uniform spheres, and lastly appearing as indefinite specks. Medullated nerve-fibres first lose their specialised substance of Schwann ; this coagulates, disintegrates, and vanishes, and the more fundamental axis-cylinder is only subsequently destroyed. It may here be mentioned that the fibres of the white substance of the cord are first laid down in the embryo as non-medullated fibres, the sheath of Schwann being of later development.

The several systems of fibres in the spinal cord, as the root-zones and cerebellar tracts, cannot be said to show degeneration in the order of their developmental rank. The posterior root-zones are developed earlier than the cerebellar tracts, or columns of Goll ; but, as we have seen, in tabes dorsalis the disease begins in the first-named segments. Unlike the units of composition in fibres and cells, the several specialised divisions of the cord are exposed to widely unequal amounts of disintegrative energy ; and this decides the order of the changes of particular segments.

Some slight indications are discernible from the physiological side that accessory functions and structures resist the invasions of pathogenic agencies with relative feebleness. At first there is ataxia only when muscular actions are not aided by vision ; but as the disease advances the assistance of the eyes in preserving equilibrium of posture or movement is of little use. Ataxia is first shown in the more complicated movements, as in turning round, or 'picking one's way,' while simple progression may be performed less erratically. And when there is ataxia of the upper extremities, the later-developed complex movements, as those required in playing the piano or in writing, are disordered before the more simple movements of gesture or prehension.

The rest of the diseases of the spinal cord, and those of the encephalon as well, will be found to furnish further illustrations. The spinal lesions of lateral sclerosis, simple, or joined with muscular atrophy, and of multiple and marginal sclerosis, are in intimate nature like the spinal lesions of locomotor ataxia. A constant and fundamental feature is dis-integration of nerve-substance and hyperplasia of neuroglia, with subordinate differences of histological detail in different diseases and cases. But, seldom is it possible, in the present state of knowledge, to say what in these leucomyelopathies the destroying forces are.

When we pass to the poliomyelopathies, whether the morbid process is degenerative or inflammatory, we find the most general description of it in the formula of dissolution. And each disease bears witness that accessory functions and structures are relatively unstable. Beautifully elaborate and incisive verifications of this truth have been unfolded by Dr. Ross.[1] In poliomyelitis anterior acuta, owing partly to the severity of the morbid process, both fundamental and accessory ganglion cells in the spinal cord are injured. It is, however, as in chronic nerve degenerations, the spinal and muscular paren-

---

[1] *A Treatise on the Diseases of the Nervous System.* By James Ross, M.D., LL.D. See §§ 371, 388, vol. ii.

chymatous elements which suffer. And, following the death of these, the neuroglia is sometimes found to be increased.

The myopathic changes of progressive muscular atrophy are a fatty, granular, and in some cases gelatinous, degeneration of the muscular bundles. The beginning of disintegration is often shown by dichotomous or trichotomous division of the fibres. Simultaneously, or, as some investigators think, antecedently, the interstitial substance of the sarcous elements augments. There appears to be no reason for excluding these changes from the general interpretation adopted for similiar affections of other organs ; namely, to regard the degeneration of parenchyma as the primary lesion, and the hypertrophy of connective tissue as consequent upon the removal of the resistance to its growth which the parenchyma naturally offers.

The neuropathic changes of progressive muscular atrophy are, again, essentially in the proper nerve matter. And here the implication of the elements according to their position in the evolutional hierarchy of tissues finds an unexpected attestation. Of the several groups of motor nuclei in the anterior horns, Dr. Ross thinks that those occupying the medio-lateral area are, in the lumbar and dorsal divisions, concerned in maintenance of the erect position, and are, therefore, evolutionally considered accessory. This group is disposed to be affected in progressive muscular atrophy. Those nuclei that occupy the median area are found in the cervical division to have grown by marginal increase, and the marginal cells are thought to 'represent a complication on the previous structure of the cord corresponding to the complication of muscular adjustments which distinguishes the hand of man from the anterior extremity of animals.' These also are most liable to be implicated when the morbid process involves the cervical anterior horns. And when there is an extension of the disease to the motor nuclei of the medulla oblongata, with the symptoms of labio-glosso-laryngeal paralysis, there is a nearly uniform disorder and extinction of functions according to their grade. Recognise, in the first place, the general fact that in progressive bulbar paralysis the affection is peculiarly one of functions and structures of high specialisation and of comparatively recent appearance in man's evolutional history. The organs afflicted are the tongue, lips, larynx, and lower facial muscles, those concerned in speech and its auxiliary, facial expression. At the onset of the disease, the more complicated actions of the lips and tongue are executed with difficulty, and finally become impossible ; and the like then happens with the simpler actions. Articulation is lost before phonation. And similarly with the complicated movements of deglutition : the final degradation of this function is seen when the simple vermicular movements of the œsophagus are paralysed. If the disease ends fatally, the most fundamental functions of respiration and circulation are in turn abolished. On the anatomical side it has been ascertained that the disease is a degenerative atrophy of motor cells on the floor of the fourth ventricle. The change commences in the hypoglossal and accessory hypoglossal nuclei, and in the nuclei of the glosso-pharyngeal ;

advancing to the facial and its accessory nuclei, to the pneumogastric, spinal accessory, and trigeminus.

Though the causation of progressive muscular atrophy and bulbar paralysis is yet *sub judice*, much may be said to the effect that energy absorbed by way of nerve channels is the exciting condition. Excessive muscular exertion is an acknowledged factor of the former disease, and immoderate use of the voice and organs of speech plays some part in producing the latter.

May the morbid anatomy of pseudo-hypertrophic paralysis be interpreted in like manner ? The alterations in the muscles in this affection are thus given by Dr. Ross : [1] 'The first muscular change which takes place in this disease consists of an increase of the connective tissue which separates the muscular bundles from one another, so that the sheaths of the muscular bundles become greatly thickened. There is also a corresponding increase of the connective tissue which passes between the fibres themselves. . . . In this early stage the muscular fibres themselves do not appear to undergo any very manifest changes, except that, according to Duchenne, their transverse striation becomes fainter, while the longitudinal striation becomes more marked. The transverse striation is, however, generally quite distinct until a late period of the disease. Duchenne regarded the proliferation of the connective tissue as the chief cause of the increased size of the muscle ; hence he called the disease '*paralysie myosclérosique*' ; but other authors believe that the muscle does not increase much in volume until the second stage of

[1] *Op. cit.* vol. ii. p. 201. Ed. 1882.

the change occurs. This stage consists of the development of fat cells in the connective tissue, and also in the newly-formed fibrous tissue, whereby the muscular fibres become widely separated from one another. The muscular fibres now become atrophied and begin to disappear. . . . After a time the muscular fibres and the newly-formed fibroid tissue completely disappear, and the entire muscle is represented by fat cells, like those of an ordinary lipoma. The fat may subsequently become absorbed, and connective tissue, with perhaps a few traces of muscular fibres, is all that is left.'

Whatever view we endorse of the pathological sequences in the muscles in pseudo-hypertrophic paralysis, the principle of dissolution receives ratification. The distinguishing substance of the muscular tissue is the first to disintegrate and die ; and in dying the organs as a whole become less complex, less special, and less multiform in structure and function. But on every ground the sequences may, I think, be most truthfully read as follows :—The sarcous elements, being the most perishable, undergo simple atrophic dissolution ; thereupon, while general nutrition is good, the interstitial tissue grows out to repair the loss, and, fat cells developing, there is acquired the semblance of muscular hypertrophy. If the atrophic dissolution continues and nutrition falls low, the fat then becomes absorbed and the interstitial substance itself atrophies, the remnants being, as just seen, strands of connective tissue and traces of muscular fibre. Did the changes begin with connective-tissue hyperplasia, then the disease would be inflammatory, as interstitial growth is commonly understood. But inflammation is disintegrative, whereas hyperplasia is integrative (chap. i. § 2, and chap. ii. § 1). Moreover, pseudo-hypertrophic paralysis does not present the clinical features of an inflammatory affection, and is not generally believed to be of this nature.

It remains to be said that the corollary from dissolution finds further general confirmation in the order in which the muscles become paralysed. The disease begins in the lower limbs, especially involving the gastrocnemii, and in the erectors of the spine. These are the muscles which have been developed during the transition from the quadrumanous to the bimanous type. It is remarkable how simian is the appearance of one afflicted with this disease when attaining the erect posture or making an ascent.[1] In the same relation it is also noteworthy that, with an inconsiderable number of exceptions, the malady is confined to children about the age when the first attempts to walk are being made. The suggestion may be ventured that a congenital or acquired feebleness of the parenchyma of the muscles affected causes it to succumb to the disintegrative energy received during functional activity. Since it remains an undetermined point whether there are neuropathic changes corresponding to the myopathic changes, reference may be omitted to the observations which have been made of lesions in the anterior horns.

Other diseases of the spinal cord supply a store of

[1] See Dr. Ross's *Diseases of the Nervous System*, vol. ii. Plate III., 1. E. 1882.

facts that may be construed by the principles of dissolution and evolution. Each affection is in union with every other in so far that functions and structures become relatively indefinite, incoherent, and homogeneous. And in a large proportion of cases these changes may be seen to proceed from the action of disintegrating forces. The many forms of myelitis— acute diffused, transverse, disseminated, chronic, &c.— bear as cogent testimony to the verity of the formulas as any of the diseases we have considered. No better example is afforded of interstitial hyperplasia as the simple consequence of parenchymatous loss than that of multiple neuritis, whatever the cause may happen to be. The corollary from dissolution that later-evolved attributes are relatively vulnerable meets with many confirmations. The order in which the muscles are affected in acute ascending, or Landry's paralysis, is one that may be mentioned. This order, according to Landry, and quoted by Dr. Ross, is as follows : ' (1) The muscles which move the toes and foot ; then the posterior muscles of the thigh and pelvis ; and, lastly, the anterior and internal muscles of the thigh. (2) The muscles which move the fingers ; those which move the hand, and the arm upon the scapula ; and, lastly, the muscles which move the forearm upon the arm. (3) The muscles of the trunk. (4) The muscles of respiration ; then those of the tongue, pharynx, and œsophagus.'

Pachymeningitis and leptomeningitis display the phenomena of serous inflammation, in which matter is abundantly set free from the blood, often causing compression of the cord. It would appear to be obligatory to clearly separate this form of secondary injury by mechanical compression from the supposed inroads upon the parenchyma by the connective-tissue growths of ordinary scleroses. The serous and fibrinous exudates of serous inflammation after causing compression may be invaded by fibro-blasts from the contiguous fibrous texture, and ultimately give place to a cicatricial growth (see chap. iii. p. 43). In this wise there is produced an appearance of injury to parenchyma by the encroachment upon it of an hyperplasic connective tissue. But, consistently with foregoing principles, the exudation of serum and fibrine is an effect of inflammatory dissolution, while the growth of connective substance is reparative and evolutional.

Since the morbid histology of the brain repeats that of the spinal cord, it would be superfluous to inquire whether in their morphological aspects diseases of the encephalon answer to the formulas. In their physiological aspects many of these diseases have been expounded by Dr. Hughlings Jackson, Dr. Ross, and Dr. Mercier, from the standpoint of the corollary from dissolution. To their writings, then, the reader is referred. One word only may be said. The so-called functional nervous affections, the diseases of the encephalon, those of the encephalo-spinal system—as paralysis agitans, hydrophobia, tetanus, &c.—and the toxic, febrile, and post-febrile nervous disorders, will be found on examination to show, not more inconclusively than previously-noticed diseases, that they are all one in being disintegrations of matter from concomitant absorptions of energy, and accompanied by transformations of function and structure of the dissolutional type.

We shall glance next at the fevers. In concluding examples in the present category of pathological changes, it may be remarked that the morbid anatomy of those organs and tissues whose diseases have not yet been considered shows the traits of dissolution and the growth of connective tissue to repair parenchymatous loss. Myomalacia cordis from disease of the coronary arteries is one of the most unequivocal illustrations. One of interest is sclerosis of the stomach where degeneration of the gastric tubules s⁺arts at the *bottom* of the glands—in the cells last developed. Others are seen in myocarditis, hyperplasia of the spleen, fibrous hyperplasia of lymphatic glands, inflammations of mucous membranes, cirrhosis of the pancreas, fibroid degeneration of the suprarenals, &c.

§ 2. The Fevers.

Some of the more characteristic structural lesions in the fevers are incisive instances of dissolution. Consider those implicating the skin in small-pox. The papular, vesicular, and pustular stages of the eruption do but mark the hyperæmic, exudative, and suppurative stages of inflammation, which have been seen to correspond with the terms of the formula. The epidermis, with its different specialised and co-operative parts, the stratum corneum, stratum lucidum, stratum Malpighii, tactile corpuscles, sudoriferous ducts, vessels, and nerves, is brought to the grade of relatively unspecialised, homogeneous, unorganised pus. And the corium, with the cutaneous appendages, sweat-glands and hair-follicles, may be involved in these changes.

Consider, again, the peculiar lesions in the intestines in typhoid fever. Peyer's patches are at first swollen and infiltrated with inflammatory leucocytes. The adenoid tissue and capillary network mainly composing these lymph follicles, the intestinal mucous membrane covering them, the submucous tissue, and even the muscular coat of the intestines, may all be converted into a uniform incoherent mass of simple round cells with dilated blood-vessels. At an advanced stage of the morbid process this mass forms a slough, the dissolution being still more decided.

In yellow fever so profound and extensive are the structural disturbances that *ante-mortem* decomposition may almost be said to occur in severe cases. Ordinary putrefactive changes appear, indeed, to set in before death. The blood is infiltrated with biliary and urinary matters ; the heart suffers acute fatty degeneration, and its substance is permeated with a brownish yellow substance ; the liver is yellow and fatty ; the lungs are congested, and the kidneys present the dissolutional changes of nephritis. And with the remaining diseases of this class—diphtheria, scarlet fever, chicken-pox, cow-pox, dengue, erysipelas, pyæmia, &c., it could be shown, were it needful, that when the morphological changes are estimable the histological elements are disorganised and made less determinate and less different.

N

Of the functional changes, those which the fevers have in common with one another may be noticed more particularly. These are perturbations of the body temperature, and of the cardiac and respiratory functions ; rigors, sleeplessness, delirium, subsultus, and involuntary evacuations. In respect of these several phenomena, the organism's functions are certainly altered as described by the formula of dissolution. Normally the diurnal mean temperature of the body is maintained with remarkable constancy under changing external conditions. The range of variation in health is probably seldom greater than 2·3° F., and the fluctuations are definitely related to rest, exercise, the ingestion of food, the hour of the day, &c. But in febrile states the temperature becomes less stable, the elevations and depressions being extreme—6° or more—and marked by an irregularity and uncertainty not observed in health. Thus the actions of the mechanisms by which a nearly uniform temperature is preserved become relatively incoherent and indefinite. And the changes in the pulse and respirations, and in the relations of these to one another and to the temperature, may be formulated in the same terms.

In rigors we see how the temperature sense becomes indefinite, the feeling of chilliness being an accompaniment of raised internal temperature and independent of external temperature. Also, that in the sleeplessness of fever the regular and specialised rhythms of sleeping and waking become quite perverted, i.e., changed to irregularity and unspeciality. Subsultus tendinum shows how the heterogeneous,

specialised, and precise actions of the healthy voluntary neuro-muscular system give place to the purposeless, inexact, and simple movements of involition.
And the indecisive, disconnected, vague, rambling,
but withal relatively simple, operations of the mind in
delirium are of the same order of functional changes.
Note how in typhus fever the mental faculties lose
their sharpness of expression, so becoming less determinate ; there is a vacancy in the eyes and heaviness
in the face. Questions are answered slowly, and
ideas are confused, showing a laboured and defective
co-ordination of the elements of thought. The special senses—taste, smell, hearing, vision, and touch—
become obtuse, and in the case of the two first-
named annulled. Thus the functions of the organ of
mind are rendered less different. Then there is the
pathognomonic feature of typhus—prostration of
voluntary and involuntary muscles. Hence result
temporary, partial, and complete dissolutions of function, such as total inability to move, constipation, and
retention of urine.

Contemplate also some of the physiological changes
of diphtheria, as when asphyxia is caused by the production of false membrane in the larynx. Then the
normal regularly recurring respirations in determinate ratio with the cardiac movements give place to
laboured efforts but little integrated with the actions
of the heart. If the laryngeal obstruction leads to
general convulsions, we have more phenomena of the
same general order. And there is complete functional
dissolution when sphincters are relaxed and reflex
excitability is abolished. The laryngeal obstruction

being fatal, the respirations become feeble and more
irregular, and finally cease, the heart continuing for
a while to beat. Circulation is now completely disunited from respiration. It may be concluded that
changes of the same nature are simultaneously taking
place in the finest chemistry of pulmonary and intra-
cellular respiration, and in the mechanical factors of
blood-circulation.

By reason of the organs involved, the secondary
affections of the nervous system in diphtheria are so
convincingly in harmony with dissolution that it may
be well to draw attention to them. Paralysis of the
pharynx and palatine arch induces dysphagia, cough,
and choking. The voice loses its individual quality ;
speech becomes less distinct and stammering from
implication of the tongue, lips, and cheeks. The
features become expressionless, the saliva dribbles,
the eyelids droop, and the head cannot be supported
by the muscles of the neck. There may be presby-
opia, myopia, and diplopia ; or paralysis of the limbs,
the heart, muscles of the chest and diaphragm ; paralysis of the bladder, with dribbling of the urine ; or
of the intestines, with constipation. Finally, all the
senses may be affected. Incontestably, these phenomena have the common trait that they are changes
to indefiniteness, incoherence, and homogeneity.

That the essence of the fevers is the essence of
dissolution—disintegrations of matter caused by the
absorptions of energy—there is no lack of proof. The
pitting of small-pox—vacant spaces remaining in the
skin after subsidence of the specific eruption—is obtrusive evidence of local destructions of tissue, by

which portions of the body are wholly severed from
it. When pitting is not a sequel, there are still the
disintegrations of inflammation passing into resolution and repair. Small-pox is often complicated by
abscesses, hæmorrhages, thrombosis, and other disintegrations. In typhoid fever, the substance of a
Peyer's patch may be so thoroughly broken down
that ulceration and perforation of the bowel ensue.
Albuminuria, of typhoid and other fevers, implies
molecular or coarse disintegration of the substance
of the kidney, and disruption of the blood. And the
granular degeneration of cells in fevers is a rending
apart of cell-substance. The examples of material
destruction in yellow fever are numerous and striking.
It is an aspect of all the phenomena. First, there is
the evidence of excessive disintegration found in the
urine. This excretion, when it is not suppressed,
usually contains an inordinate quantity of urea,
'reaching as much as a thousand grains per diem.'
The average in health is 512·4 grains. And in this
disease, the excretion of an ounce of albumen in
twenty-four hours has been observed. Renal disintegrations are represented by cells of epithelium,
tube-casts, and *débris* in the urine. Black vomit
acquires its notorious quality from abnormal disintegrations—blood set free from the gastro-intestinal
mucous membrane. Black vomit also contains epithelial cells and urea. In yellow fever, hæmorrhage
may take place from the gums, eyes, and ears. The
fibrine of the blood is diminished—presumably because it is destroyed—and there is said to be an
unwonted dissolution of red blood-cells. Desqua-

mations occur in the kidneys, and the heart and liver evidence those chemical disintegrations of cell-albumen which terminate as fatty degeneration. And the nervous phenomena—coma, convulsions, paralysis, &c.—of yellow and other fevers are almost certainly the outcome of molecular and molar demolitions of tissue.

Finally we come to the disintegrations which are supposed to signify increased tissue oxidation. In recognising the fact that the febrile state is attended with an enlarged discharge of the products of tissue metabolism, we are not committed to any hypothesis as to the chemical and physical sources of these products, or as to the causes of the heat of fever. Whether the pyrexial state is owing to increased oxidation, or to the liberation of heat that would in the normal state have taken other forms of energy [1] —perhaps heat or chemical energy—or whether we must look for an explanation to the actions of the nervous system, is beside our present purpose. That high temperature is invariably accompanied by tissue dissociations does not admit of doubt. The carbonic acid exhaled by the lungs and the nitrogenous excreta of the urine are always augmented in fevers. Fevers, too, are accompanied by visible wasting, showing that the involved processes are destructive.

Are these processes the consequences of absorbed environmental energy? Less unobviously so, perhaps, than the morbid phenomena of any other class

[1] See 'An Introductory Address on the Heat of Fever,' by W. Ord, M.D., F.R.C.P., *British Medical Journal*, October 24, 1885.

of diseases. Small-pox, cow-pox, and vaccinia no hesitancy is felt in ascribing to the entrance to the blood and tissues, through contagion or inoculation, of some form of virus ; and we believe that the multiplication of this within the body sets up the chain of disintegrations constituting the morbid anatomy and physiology of these affections. Facts are thought to warrant the view that a poisonous effluvium, elaborated in the overcrowding of human beings, is the determining circumstance of typhus fever. It may be regarded as almost established that typhoid fever has its source in matter which gains access to the organism by way of the alimentary canal. The energy so imbibed causes the disintegrations of intestinal catarrh and inflammation of the lymph follicles, and, by continuity of tissue, its influence is passed on to the mesenteric glands, blood, and other structures. It is supposed by several investigators that the active agent is a bacillus. Of scarlet-fever, and all diseases allied to it by nature, there is now a firmly grounded conviction that their causes exist in external conditioning forces. Their communicability from person to person imposes this conclusion. What the concrete factors are has yet in many cases to be determined. But the goal of research in this field is perhaps well indicated by the recent investigations of Dr. Klein, Mr. Wynter Blyth, and Mr. Power upon the connection of scarlet fever with the inception of a micrococcus.

The fevers show us as other diseases do the intervolution of dissolution and evolution. Though we cannot penetrate to all the details of those processes by which the general equilibrium of the func-' tions is restored in the acute specific diseases, yet the processes are of the same order as resolution and repair. They are a reversion of the previous dissolution, since the lost qualities are recovered. In the poison-diseases, like scarlet fever and diphtheria, recovery must be largely effected by gradual elimination of the poison, and by the redintegration of inflammatory exudates. The focal lesions of typhoid fever supply good examples of the reparative changes in that disease. A solitary or Peyer's gland may undergo resolution, in which event the hyperæmia subsides. The inflammatory extravasations are then absorbed or reaggregated round common centres, matter being integrated and motion lost (chap. ii. § 1). And the gland re-acquires its normal organisation, distinctness, and variety of functions and structures.

If the morbid changes end in necrosis, there results from rupture of the follicles a typhoid ulcer. At the site of this repair takes place ; there is a reproduction of cicatricial tissue strictly analogous to the fibrosis of chronic diseases, and, as in those diseases, the higher structures are perceived to be weaker than the lower (cor. 1). Peyer's patches are differentiations of the intestinal lymphatics, are accessory elements in the intestinal mucous membrane.

The changes in the kidneys in scarlet fever and in yellow fever are the verifications of the corollary we observed in the section on Bright's disease. Others may be followed out in the functional and nervous phenomena of febrile diseases. Speaking broadly, we obtain glimpses of a conformity to the principle in the invasion of the neuro-muscular and other functional systems, and in the invasion of the mental faculties in fever, especially in typhus fever.

To lay hold of the fact that the process of equilibration in acute disease is the opposite of dissolution, dissolution may be contemplated from the dynamic aspect. From this single point of view it is an increase in the relative movements of parts and a decrease in the relative movements of wholes. There is a regression from the motions of large masses to the motions of smaller masses, and from the motions of small masses to the motions of compound molecules, and from the motions of compound molecules to the motions of simple molecules. We reverse the terms, of course, for evolution. Then 'evolution is a decrease in the relative movements of parts and an increase in the relative movements of wholes—using the words parts and wholes in the most general senses. The advance is from the motions of simple molecules to the motions of compound molecules, from molecular motions to the motions of masses, and from the motions of smaller masses to the motions of larger masses.' A case of typhoid fever will illustrate the general dynamic changes so exhibited. The visible movements of large and small wholes, as the movements of the body and viscera, are derived immediately from the molecular motions of the involuntary and voluntary muscles, which molecular motions are in turn remotely and partly derived from food. Therefore, by the terms given above, in dissolution the motions should be given back in this order: the large wholes should yield up their motions to small wholes, and the small wholes their motions to smaller wholes, and these their motions to molecules of descending orders of complexity. Such transference can plainly be traced. In typhoid fever we find an increase in the relative movements of the parts and a decrease in the relative movements of the whole. If the molar movements of the body—the movements of the body and limbs through space—are not altogether arrested, they are notably diminished, and, conformably, they appear as added motions of parts. First to be noticed is the increased molecular activity implied by the raised temperature, and that in proportion to the height of the temperature there is indisposition to voluntary movement. Next, the heart, diaphragm, and subsidiary organs of respiration perform their respective rhythmical motions more rapidly ; there are increased movements in the component parts of the spleen and intestinal lymphatics, and of the intestines, as evidenced by enlargement of the former and the more active contractions of the latter. Then certain structural components of the bowel, in the shape of the solitary and Peyer's glands, undergo such augmentation of their molecular momenta that the histological particles are disunited and acquire independent motions, leaving areas of complete dissolution—typhoid ulcers. If we followed these small disintegrated elements in their subsequent careers we should find their individual motions resolved into the motions of the smaller units of their composition, and the motions of these resolved into the motions of their compound and simple molecules. With death as the termination, the motions of the organic systems— cardiac, pulmonary, secretory—gradually abate, and in the end vanish into the invisible motions of dead organic matter, the succeeding changes of decomposition answering to the formula.

Since recovery from acute disease implies a re-integration of matter and motion, convalescence from typhoid fever will show the evolutional redistribution of motion. The earliest premonition of recovery is a depression of the exalted molecular motions—heat vibrations. Coincidently the organic functions— respiration and circulation—acquire slower rhythms. By increased assimilation, the molecular energy contained in alimentary substances is stored up in the tissues and made over to the organs. These smaller motions are then compounded into the molar motions of the individual, and so finally dissipated. It must be admitted that the changes of motion in disease are thus exhibited in an extremely general way. Perhaps this is as much as can be expected where the involved forces are so variously and definitely relationed as in the human organism.

We shall now take examples of diseases which may involve the entire system, that are not contagious or infectious, and not necessarily attended with coarse lesions. These affections begin as functional disorders, and therefore the structural changes are molecular rearrangements.   Diabetes will serve to represent the class.

§ 3. DIABETES AND ALLIED AFFECTIONS.

The facts of diabetes, and the disorders to be regarded as its congeners, gout and rheumatism, we shall find amenable to synthesis by dissolution and evolution ; and when their affinities of nature and causation have been laid bare, our notions of these, as of other diseases which have been brought under inspection, will, I think, grow in vividness and consonance with fact from the influence of the formulas.

If we have in mind average, not exceptional, examples, i.e., ordinary cases of diabetes, the morbid anatomy is very indefinite.   In its first stages we are obliged to consider the affection a functional one. But such positive changes as *post-mortem* examinations have revealed, and those to be viewed as secondary and complicating lesions, exhibit no discrepancy with dissolution.   After the repeated illustrations that have been made the mere nomination of them will suffice.

In a moiety of cases the liver has been found hyperæmic, and in advanced cases with the signs of atrophy ; parallel changes have been discovered also in the kidneys.   The pancreas was small, hard, and bloodless in thirteen out of thirty cases reported by Rokitansky.   The tongue is commonly smooth and red, and its epithelium is atrophied.   Phthisis and pneumonia are very frequent secondary affections, as are boils, carbuncles, and cataract.   Morphological changes in the nervous system have been described by Dr. Dickinson. These anatomical characters of diabetes, though inconstant and not free from ambiguity, do at any rate appear to be dissolutional.

From the clinical and physiological sides varied and tangible abnormalities are presented. If we take first the most important, the deliverance through the kidneys of an excess of saccharine matter, we have to learn whether this phenomenon is assimilated by the formula of dissolution. Is the increased excretion of sugar a reduction to relative simplicity, generality, and uniformity of functions and structure? It must be allowed to mark a change of this nature both in the constitution of the urine and in those physiological actions of which the constitution of the urine is an outward and scrutable sign. When we dealt with the facts of Bright's disease it was remarked that normal urine has a definite, coherent, and heterogeneous organisation ; its composition is constant, stable, and complex. And the observation was made that urea in human urine being the last and in a sense highest term in the series of chemical changes which food undergoes in the system, the presence in the urine of bodies intermediate between urea and the crude proteids and carbo-hydrates of the ingesta renders the structure of the urine relatively simple and general. From the dissolutional point of view, we must so regard the excretion in excess of the common urinary constituents—oxalic acid, uric acid, urates, phosphates, &c.—and *a fortiori* the excretion of albumen, sugar, or fat. And if the urine is thus changed, so must be the metabolic processes of which the quality of the urine is representative. Therefore, the presence in diabetes of an abnormally large quantity of sugar makes the urine less complex and less special than it would have been had such sugar, taken in as aliment, been elaborated by the tissues and the residue cast out as normal excrement. If the sugar may be supposed to take the place of different byeproducts of its conversion into animal energies, then the urine, by the presence of the sugar, would be to that extent rendered less heterogeneous. And this abnormal excretion of sugar implies changes of identical character in the metabolic processes, whatever the internal or external cause of these changes, whether a tumour, injury to the brain, or the ordinary cause of diabetes.

It seems proper, then, to regard the leading feature of diabetes as in agreement with dissolution. Other symptoms of the disease as well may be comprehended under this formula, namely, diarrhœa or constipation, excessive urination, enuresis, the loss of sexual power, coma, and failure of the mind. But it will be well to pass from these to correspondences with dissolution of greater interest.

Are the dissolutional changes of diabetes disintegrations of matter from the taking up of energy? The question is unanswerable unless we know the causes of the disease ; but pathologists are as uncertain upon the ætiology as upon the pathology of diabetes. What, then, must be said? The following hypothesis of its causation appears to me the only tenable one.

In the search for causes the principle of dissolution ever conducts us to the interactions of the organ-

ism and its environment. Fixing attention, then,
first upon the external conditions of the untreated
diabetic, we are arrested by his immoderate ingestion
of amylaceous, saccharine, and other articles of food.
Here, I think, lies the prime external factor. The
internal change is the failure to convert aliment into
the normal forms of animal energy and waste matter.
What in the anatomical and physiological senses the
fault exactly is we cannot say ; but about the fact there
is no doubt. Possibly the change is in the general
protoplasm of the body. All the circumstances in the
lives of persons suffering from confirmed glycosuria
favour, I think, the conclusion that from the long-continued ingestion of food exceeding in quantity the
needs and assimilative capacity of the body, there has
followed an exhaustion of those forces through which
food is converted into the higher forms of energy-
yielding animal substances. The needs of the organism are, of course, relative to various physiological
states, and thus the excess may be a relative excess.
But it is as we see in other diseases, that undue functional activity—in the present case the activity of the
function of elaborating integrable food-material—leads
to the abolition of function, or functional atrophy.
A condition constant in all forms of atrophy is the
predominance of waste over repair, the waste being
related to energy absorbed (chap. iii. § 4). In
diabetes, then, the food taken in is the energy absorbed ; this wears out the organs concerned in the
elaboration of the carbo-hydrates of food, just as the
immoderate ingestion of proteids will wear out the
kidneys. Some forms of albuminuria may, I think,

be compared with ordinary diabetes in respect of
causation and the fact that, crude food-material circulating in the system in excess, the excretory organs
become the channels of discharge.

Upon this understanding of the ætiology of diabetes, we readily make an arrangement of its phenomena as they are naturally related.   Early in
the history of any case, the appetite, in health an in-
dication of physiological needs, becomes perverted.
Growing by what it feeds upon, the desire for hydro-
carbons soon ceases to bear a proper relation to the
physiological wants and powers of the system, and
appetite becomes a craving.   Ministrations to this
craving lead at length to the undetermined disintegrations of function and structure which are the
anatomical and physiological bases of confirmed diabetes.   There results inability to deal with even a due
amount of hydro-carbonaceous food.   An excessive
quantity of sugar is thus thrown upon the blood.
This sugar naturally seeks an outlet by the kidneys ;
and if these organs are unequal to the work of excretion
the sugar will find other outlets.   The diarrhœa not
infrequently associated with diabetes signifies efforts
at elimination by the bowels.   There is probably a
determination of saccharine matter to the skin.   The
inordinate thirst of diabetes is a physiological adaptation which leads to an increased ingestion of water,
and so to the solution and discharge of the sugar.
Hence the polyuria.   In large part, the remaining
symptoms of diabetes are connected with disintegrations produced by the blood impurities ; the skin
affections, emaciation, exhaustion, and diabetic coma

may be so explained. It is very interesting to observe in this disease a corroboration of the opinion expressed when we considered the causes of pneumonia and phthisis; namely, that the quality of the blood of the pulmonary artery as related to alimentation is an important determinant of certain diseases of the pulmonary tissue. A very common termination of diabetes is by pneumonia and phthisis. We speak of diabetic consumption. It may be concluded that the blood which passes to the lungs from the liver and general system, laden with food-material not prepared for the lung's reception of it, is one of the causes of the disintegrations of diabetic pulmonary disease.

As we discriminate those changes in diabetes that are evolutional in character we are reminded that the most approved treatment for the disease is in entire concurrence with the view here taken of its causation. By appropriate measures, which include a restricted consumption of starchy and saccharine matter, partial, and even complete, recovery may be effected. Hypertrophy of the bladder is a subordinate evolutional change of interest at the present moment. This concomitant of diabetes has only lately been made known to the profession by Dr. Robert Maguire. Clearly, the growth of vesical tissues is consequent on increase of function (diuresis).

Though certain of the individual phenomena of diabetes exemplify the disappearance of functions in the inverse order of their development, the essential lesion involves a fundamental function. It is probable that the consumption of 'free' sugar as an article of food—a custom which later civilisation has made possible—is a potent cause of diabetes.

Allied diseases, as gout and rheumatism, may be similarly interpreted. There is now a general agreement among physicians that rheumatism and gout are due to the presence of a poison in the blood, and that the two diseases are closely related in nature and ætiology. But their affinities with diabetes are less generally recognised.[1] The poison of gout (uric acid) many now believe to be generated by a diet too rich in nitrogenous elements. So that in gout as in diabetes we must consider the ingested aliment as the most important external cause, and the failure to deal physiologically with this aliment as the necessary internal defect.

Rheumatism probably differs only in the chemical characters of its poison, in certain secondary conditions of its production, and in the distribution of it in the system.

The reader will be prepared to comprehend their phenomena from our special points of view, and also the phenomena of kindred affections, as scurvy and purpura. A concluding word may be said to point out that the appendicular and accessory structures, the cardiac valves, are peculiarly susceptible to the action of the poisons of gout and rheumatism, a fact which falls in with the corollary from dissolution.

---

[1] See *The Croonian Lectures on Some Points in the Pathology of Rheumatism, Gout, and Diabetes.* By P. W. Latham, M.A., M.D., F.R.C.P.

§ 4. DISEASES OF THE MIND.

Are affections of the mind to be given a position in pathology that will mark them off as requiring presentment essentially different from that we have employed for other diseases ? Though the subject-matter of subjective psychology comes before the investigator in a form absolutely unique, it is believed that the facts of mental pathology may be dealt with on precisely the lines adopted for the pathology of the body. Psychical activity is here assumed to be the equivalent of cerebral activity. As contraction is the property of muscle, secretion the property of a gland, and the nervous discharge a property of nerve tissue, so is it postulated that feeling, knowing, and willing are properties of cerebral substance. Mind and body are two aspects of the same thing, as motion and matter are two aspects of the same thing. The proposition will not be understood to identify these properties. On this ground, which, I think, few alienists will disallow, anatomical changes form the physical bases of abnormal mental changes. Cases of insanity, however, singularly seldom supply demonstrable proof that psychical disorders are attended with corresponding structural alterations. By far the larger number of diseases of the mind have not yet been associated by pathologists with a visible morbid anatomy. Dementia, mania, melancholia, delusional and other insanities, afflicting perhaps the bulk of asylum patients, are, on the physical side of their phenomena, little understood. Many of these probably fall into the category of what are known as functional diseases; *molecular* rearrangements are their physical counterparts. Of the residue, the morphology remains to be discovered. There are, nevertheless, numerous varieties of mental disorder which have a distinct morbid anatomy; and to one of these we may at once pass.

In setting out upon our consideration of diseases of the mind, the affection known as general paralysis of the insane may be chosen as the point of departure. If we first place the chief facts of this common and well-known malady in juxtaposition with the principles the verity of which we are testing, other forms of insanity can afterwards readily be included in the necessarily rapid *aperçu*.

We have to learn whether the sensible bodily changes of general paralysis are the changes predicated by dissolution. They are in every material respect those we have encountered again and again in the analysis of other diseases—disintegrations of the substance of the body and a reduction to a deranged, obscure, and general configuration, with integrations of inferior order. If we study the minute characters of the principal lesions of the nervous system, what do we find? That nerve-fibres both of brain and spinal cord are without medullary sheath or have been bodily destroyed; that nerve-cells have withered or have passed wholly out of being. There are also the signs of inflammatory disintegrations. Leucocytes and fluids are seen outside their normal areas of congregation. Vessels are degenerated. By the side of these structural transmutations we find the products of formative processes—false membranes from pachymeningitis and a new growth of connective tissue (see p. 175). Often there is cirrhosis of the kidneys. In this disease as in other organic diseases, the root-change is a dissolution of the specific elements of the functional systems involved.

And what are the physiological equivalents of these molar sensible lesions and of those co-existing molecular lesions that are beyond the range of sense-perception? Are they also uniform with the principle of dissolution? The bodily symptoms and the motor phenomena of general paralysis we may pass over, since they are reproductions of phenomena already noticed. Occupying ourselves, then, with the leading mental symptoms, let us contemplate these under the usual psychological classification which groups the operations of the mind in an ascending scale of their complexity. We begin with sensation. Do sense-impressions become less definite, coherent, and heterogeneous in general paralysis? Organic sensibility—meaning by this the effects upon consciousness of the physiological working of the stomach, kidneys, liver, &c.—is vague as compared with special sensibility—hearing, seeing, &c. Undoubtedly organic sensibility becomes more vague in general paralysis of the insane. One of the best-marked peculiarities of this affection is the insensitiveness of patients to changes in the internal organs. Profoundly destructive actions may be going on in the tissues of the nervous system, kidneys, stomach, and other viscera, yet the general sense is one of *bien être*. In still other respects organic sensibility is blunted. By the gluttony of general paralytics in the stage of dementia we are to understand that the excitations of the gastric sensory nerves are no longer represented in the sensorium in a manner enabling the individual to eat with discrimination ; and so patients surfeit themselves. Or there is no memory of a meal just eaten, and one meal is taken immediately after another. These changes imply a loose compounding of and relative uniformity in the waves of visceral nervous energy going to the abode of consciousness.

Special sensibility is not as deeply involved in general paralysis as organic sensibility ; there may, however, be local anæsthesia, loss of tactile and muscular sense, also loss of the sense of smell, and colour-blindness.

These faculties—sensations—are the unwrought substance of perception and memory. Perception is the mental reference of sensations to objects ; thus a sensation of blue is localised or externalised in the sky. Memory is the reproduction of perceptual impressions ; after turning from the sky the perception is recalled in the form of a mental image. Are perception and memory affected in general paralysis in the way dissolution sets forth ? Early in the disease the keen edge, the definiteness, of perception is merely dulled ; but later, in the stage of fatuity, the power of perceiving things is scarcely greater than that of an infant's. The sounds and sights of every-day life call out no responsive signs, and it may be impossible to bring the patient's mind into relation with distant or minute objects, to cognise which there must be a strict co-ordination and sharpness of faculties. And things that are perceived are liable to be immediately forgotten. The events of yesterday cannot be recalled. Appointments slip the memory ; the patient breaks down in conversation because he cannot remember facts and words and their relations. The composite train of images required for the recollection of a group of experiences—say the incidents of a day's pleasure excursion—are offered in fragments. The mnemonic process is marked by indistinctness and relative simplicity of the pictures and a feeble integration of them. Corresponding with the cerebral changes, memory is disintegrated.

The grandiose ideas of the insane paralytic do not indicate development but degradation of that form of memorising known as imagination—the constructive reproduction of images. The imagery in 'exaltation' is not truly constructive : there is lacking a mutual dependence of parts, and the images themselves rarely range beyond a narrow egoism.

What is the general character of thought, judgment, and reasoning ? Since it is out of perceptions and memories that concepts, judgments, and ratiocinations are built up, it might be expected that the less elementary faculties would share the changes of the more elementary. To the higher intellectual operations as to the lower, an applicable criterion is that of their clearness, coherence, and speciality. In respect of these qualities, thought, judgment, and reasoning exhibit regressive changes. The person who purchases for his own use 'a dozen broughams and twenty parrots' is one in whom there is no perception of agreement between impulses and needs. In one who states the proposition that he is about to 'marry the Queen and all the princesses,' the judgment and reasoning are insane because the relation of congruity which should subsist between desires and the possibility of satisfying them is thoroughly confused. As to the general symmetry of thought, 'incoherence' is the common term employed to denote the discontinuous and disconnected utterances in delirium and mania, two constant phases of general paralysis. In variedness or speciality there is, as in other examples of dissolution, only an apparent increase. Thought is not discursive, but is occupied almost exclusively with selfish concerns—is quite unspecialised.

Were it required, it could be shown that on the emotional and volitional sides of mind the process of dissolution can be traced. The simple feelings, the sentiments, and the will, all undergo analogous changes. This will appear as we apply to the mental symptoms of general paralysis the corollary from dissolution. We may now address ourselves to this task, and leave the causation for later notice.

Can there be discerned in the disease we have selected to typify insanity a derangement of mind the reverse of its developmental arrangement? Are the mental faculties we have just hastily viewed degraded from above downwards? They are clearly so. Insanity in its medical and medico-legal aspects engrosses attention as it affects conduct. Conduct is the complex action—always involving muscular movement of one kind or another—by which character and the relations to environment, especially social, are delineated. It is thus the sum and representative of all mental endowments, and the last of the series of attributes unfolded by development in the individual, and produced by evolution in the race. That the first indications of mental alienation are alterations in conduct is, then, quite in keeping with the corollary from dissolution. Let us now see *how* conduct is altered, and note the order of the changes in the mental and moral qualities.

The symptoms of general paralysis have been grouped by three periods of the disease : the stages of alteration, mania, and progressive dementia. This arrangement it will be convenient for us to keep in mind, though our purposes require a contemplation of the changes in continuous series.

General paralysis is early betrayed by a change in habits and manners. One who was wont to be cleanly and orderly in his attire becomes uncleanly and slovenly. The self-respecting actions and respectful bearing towards others which are distinguishing marks of the higher orders of men give place to indecencies and rudeness. The person may be exposed without regard to circumstances ; foolish assaults may be made upon women ; passionate and inconsiderate orders are given to servants. The punctilious man disregards the times of meals and appointments ; the prudent man spends money lavishly and rashly. Now these alterations of character and conduct signify a *loss of lately acquired qualities*. The manners and habits of civilised man are among the highest and latest products of human evolution ; and those

qualities which take their place in one suffering from
general paralysis are the qualities of earlier and inferior
creatures. The man is, as we say, degraded. It is
here as with dissolution everywhere : the root-change
is not a something added but a something taken away,
and the loss of the higher attributes makes active the
latent lower.

Nearly all the mental features of general paralysis
of the insane can be interpreted in this wise. Let it
be the highest form of volitional activity—say, deli-
beration upon a course that is not clear—and the
business man threatened with the disease is unequal
to his situation. He cannot be trusted to decide
upon his business affairs. Effort of will as applied
to the complex circumstances of social life he is unable to exercise ; but he may be most wilful in the
simpler and lower circumstances, as when bent upon
doing mischief in a fit of mania. The superior mental quality known as self-restraint is lost early, and
the loss can be traced through a descending series.
Continued self-restraint leads to habitual action, and
habit in moral, social, and intellectual activities decides
what we call character. As we have seen, a change
in individual character is usually the first to attract
attention. Self-control is the subordination of the less
specialised endowments to the more specialised ; and in
the sphere of control of feeling it is at first only the
more specialised feelings over which restraint is lost.
The forbearances commonly exercised in the minor
social relations disappear, to be replaced by petulance,
sulkiness, unreasonable anger, and exacting conduct
with servants and family. Then the lower desire to

obtain unlawful possession of the property of others ceases to be checked by prevision of possible penalties or by altruistic considerations ; the general paralytic will commit theft.

Attention, which involves direction and fixing of the thoughts, wanders or is unsteady. Or the attention cannot be engaged when the stage of dementia has been reached. Mania shows an absolute abrogation of the capacity to regulate thought by keeping in the mind particular trains of images. The sufferer from this malady is also bereft of the power to inhibit the more active impulses.

In the different stages of the affection, loss of control is seen over every order of movements. In the early stage there is restlessness, when patients walk and talk incessantly ; and in the final stage involuntary evacuations of the bladder and rectum. Loss of control over the sphincters, a symptom of many diseases, I believe to be the loss of a late acquisition. The social conditions of primitive man would not have required that regulated action of the bladder and rectum which existing social conditions impose.

When we turn to the feelings a like degradation is observable. The higher sentiments, as the moral and æsthetic, are the first to disappear. Lying, theft, and homicidal tendencies are the traits exhibited by the insane paralytic who was once a respected member of society. Dirtiness of person and language are substituted for decency and refinement. There is from first to last a pretty regular transitional change of feelings ; from the superior impersonal ones of sym-

pathy and affection, and sentiments that are cognate,
to the inferior egoistic ones. During the stage of
' exaltation ' a concern for family is almost lost in
inordinate self-esteem, vainglory, and other relatively
low-level traits. In the closing periods consciousness
is mainly occupied with the simplest sensations—those
of hunger and its satisfaction.

Recent acquirements are plainly shown to be the
least stable in the region of the active faculties.
' The pianist loses his skilled touch ; the actor fails
to learn a new part ; the ready salesman no longer
has his great facility of selling.' Lastly, the intellectual faculties also decay from above downwards,
and become less complex, coherent, and heterogeneous.

Now we have to see in what manner these losses
are occasioned.   Between the insanity of general
paralysis and diseases of the body we have found an
absolute unity in the broadest characters of their
phenomena.   Neither morphologically nor physiologically does paralytic insanity form an exception
to the ultimate clause of the formula of dissolution.
But is it not exceptional in its causation ?   Can the
disintegrations of mind and body be connected with
concomitant absorptions of energy ?

When we looked at the diseases of the spinal cord
it was found convenient to differentiate the paths by
which disintegrating forces reach its parenchymatous
elements. It was seen that, speaking generally, these
paths are in the blood-vessels or in nerve-tracts. We
observed how in the case of ordinary tabes dorsalis—

a disease frequently associated with general paralysis —the destruction of fibres in the posterior columns is caused probably by energy received through functional systems of nervous tissue. By what channel or channels is taken in the energy which produces the disintegrations of general paralysis? We are still wanting a satisfactory ætiology of paralytic insanity. It is approximately the general professional belief that the ordinary disease is caused by an association of conditions. · These are excesses of various kinds, and chiefly sexual and alcoholic excesses. The excitements and anxieties incidental to the struggle for wealth and social position are said to be contributory factors. We must, then, for the present recognise at least two paths by which absorbed energy reaches those elements of the brain and spinal cord, the disintegrations of which are the acknowledged lesions. One is by the blood, and the other by the nervous system. The chemical energy of ingested alcohol, on one hand, and the nervous excitements of sexual excess and the strain of life, on the other, are the recognised causal conditions. Without doubt, however, there remain many unrecognised factors. Yet, as far as existing knowledge permits us to speak, it may be said that this form of insanity is just as complete a fulfilment of dissolution as a common burn or a fractured limb. The mind elements in stratum after stratum, from periphery to centre, are extinguished by the disruptive action of incident energies. If we look at the mind as an organic substance, formed by the compounding and recompounding of simpler elements, general paralysis is its analysis.

Each step in the process of decomposition brings to view the constituents of the group decomposed ; and these constituents are the symptoms of the several stages.

All diseases of the mind are neither by ætiology, symptomatology, nor morphology as readily made one in principle with other diseases as is general paralysis. The causes of many forms of melancholia, acute mania, dementia, hysteria, &c., have yet to be made out. Nor has the decomposition of mind in all the other insanities that relatively regular character seen in the disease we have just noticed. Also, as previously remarked, our knowledge is very meagre of the physical correlatives of mental alienation in general. Yet if the concurrences with dissolution are less satisfying than in the case of diseases of the body, there is on the whole much harmony with the principles. Hysteria, hypochondriasis, and melancholia are varieties of one species. In each the lower activities assert themselves in consequence of the removal of functions by which they are normally kept in subordination. Hysteria shows how the feelings and emotions of sexual life start into prominence with the disappearance of faculties that regulate the social life. Hysterical patients of the worst kind become passionate, indecent, untruthful, dishonest, and lazy, but cease to be self-supporting. The energies of the hypochondriac are absorbed in self-contemplation and anxious concern for the health of his body ; while the melancholic lives in a state of fear or despair rather for his social, moral, or spiritual welfare.

We see the abolishment of the non-egoistic or higher egoistic of the mental endowments, and the survival of the lower. If an exhaustive examination were made of the physical and psychical phenomena of these and related affections there could be exhibited a general agreement not only with the corollary from dissolution but also with the formula. It is remarkable how consistent with our principles of interpretation are the phenomena of acute mania. It may be caused by fevers, alcoholic poisoning, or the administration of chloroform. Dr. Savage says : ' The excitement and restlessness are results of want of control, not of excess of power. The circulation is feeble and the power of reaction is small. . . . The pallor of the face points to the anæmia, and the wide pupil to nervous weakness. . . . Histologically, I have rarely failed to find in fatal cases of acute mania changes in the nerve-cells, more especially in the pyramidal layer. The nerve-processes may be wanting ; the cells may be smaller and indefinite in outline ; and in some cases they are much wasted, or suffering from degeneration of one form or another.' Dementia supplies an exemplification of dissolution almost equalling in probative value that of general paralysis ; and idiocy and imbecility are also degradations to mental simplicity, incoherence, and uniformity.

Does what we know of general ætiology in insanity authorise the statement that the causes of diseases of the mind and the causes of diseases of the body are alike in nature ? Are both due to the dominion of the forces of the environment, to the consumption of external energy ? Whatever classification is made of

the ascertained true causes, they are synthesised by
the general law. If the predispositions to insanity are
ascribed to race, sex, age, or inheritance, the factors
about which we are most certain and most interested
are definite life-circumstances to which the organism
succumbs. These are referred to as domestic trouble,
grief, sorrow, adversity, anxiety, overwork ; excitements of the social world, of religion, and of the pas-
sions ; shock, intemperance in eating and drinking,
starvation of body and of mind, idleness, accidents,
the poisons of venereal disease, the fevers, pregnancy, &c.

It only remains to add that in the region of mind-
disease dissolution and evolution are twin processes.
As in diseases of the body, inferior evolutions accompany every larger dissolution. This is exemplified
by the growth and organisation, especially among the
chronic insane, of delusions, hallucinations, and the
vices of the alienated. The gradual increase and
establishment of a delusion of persecution as the outcome of a process of dissolution is in parallelism with
the gradual increase and establishment of appetite in
diabetes, of convulsions in epilepsy, and of connective
tissue in hepatic cirrhosis. In practical medicine
these mental excrescences usually monopolise attention.

The evolutional changes of organic equilibration
obtain in madness as in other deviations from the
normal, but with special force in acute madness.

Here we reach the last of our examples ; but were
it desirable others could be drawn from every classi-

fied group of morbid actions. The detailed application of the principles to many disorders that have
been passed by without mention would, I believe, give
a new and better meaning to their phenomena and
direct us to the discovery of valuable data. Diseases
of the eye and other special organs, and of the circulatory, digestive, and reproductive systems, furnish interesting illustrations. Leucocythæmia is a notable
one of these; and among functional diseases not one is
more instructive than sea-sickness. In this affection
we are shown how simple rhythmical molar motions in
the environment are used up in effecting molecular
changes in each functional system. The oscillations
of the vessel probably first conflict with the motion
of the blood of the carotid veins, and secondarily
with the cardiac rhythms. Thence result anæmia of
the cerebral vessels and, correlatively, the psychical
disturbances of sea-sickness, the latter according re-
markably with the terms of dissolution. Low systemic arterial tension, constipation, scanty urination,
and other pathognomonic symptoms are the tertiary
phenomena. Sea-sickness and the disorders caused
by poisons afford some of the most forcible examples
of the principle of equilibration (§ 3, pt. iii.).

PART III.

IMPLICATIONS.

CHAPTER I.

THE RESULTS AND CONTINGENT PRINCIPLES.

§ 1. Summary.

WE have now followed out with some fulness the illustrations of dissolution and evolution in pathology. It is therefore incumbent to retraverse our steps and take up the results by the way, to carry them to conclusions of practical utility.

When we assembled the facts connected with the innermost changes of inflammation we found its numerous complicated essential phenomena synoptically exhibited in the brief and simple formula of dissolution with which we set out. Disintegration was defined as an increase in the centrifugal motions of the units of a structure, and it was seen that inflammation is a disintegration. Diapedesis, the emigration of leucocytes, the exudation of serum from vessels, the dilatation of vessels, the production of capillary stomata, and cellular proliferation are one and all phenomena of this nature. And the sequels of inflammation — suppuration, necrosis, sloughing, hæmorrhage, &c., were discovered to be like but more pronounced redistributions of matter. Inflammatory contraction of vessels and inflammatory stasis were referred to as integrations of matter; and, not to anticipate subsequent exposition, it was reserved for a later chapter to show that such secondary integrations are not at variance with the principles to be employed, the truth being that evolution and dissolution are indivisible processes.

It was next observed that inflammation and its consequent suppuration are, with rare exceptions, referrible to the operation of external forces. This sustained the proposition of dissolution that disintegrations of matter proceed from absorptions of energy. So it was disclosed that inflammation consists in the shaking asunder of the units of an organ or tissue by incident energies; but since inflammation is a change pertaining only to vessels it is primarily a dislocation of the units of these structures. Inflammation may be paired off with the disintegrations of inorganic aggregates.

We then went on to learn how the functional quiescence of an inflamed organ is to be accounted for, and afterwards whether inflammation and suppuration agree with dissolution in being changes from a definite, coherent heterogeneity to an indefinite, incoherent homogeneity. We found that they do. In this way the particular phenomena previously cognised as more or less separate and individual were affiliated with one another by natural cause and effect.

Resolution and repair corresponded with evolution; and when we came to consider the retrograde metamorphoses the occasion arose for emphasising a point not to be overlooked: the inseparableness of

evolution and dissolution. Attention was called
to the fact that either is a differential result of the
working of both, and that whichever result ensues
as the superordinate change there are subordinate
changes of the opposite kind. Coagulation of the
blood was shown to be a disintegration of the blood,
but the solidification of fibrine a secondary integration.

It was next brought out that the formula of dissolution sets before us with much fidelity the most
general phenomena of regressive metamorphosis and
of diseases produced by animal and vegetable parasites, the latter presenting obvious minor evolutional
changes.

The facts of infective tumours offered little opposition to unification by the formulas; but, to give
order to the phenomena of true tumour-formation,
it was found necessary to attempt to determine the
nature and origin of the true tumours. The position
was taken that tumour-cells are the agamic progeny
of normal tissues, and are thus the outcome of a
process of genesis. Many and varied groups of data
upheld this view. It was seen to be supported by
the embryology of tumours and by their histology,
and sanctioned by the general uniformity of their
phenomena with the phenomena of reproduction at
large. Observations, isolated but too plain to be
called in question, upon the causes of tumours necessitated the inference that tumour-cells or germs are
generated by the tissues in consequence of declining
tissue-vitality. Many coincidences with this hypothesis were at hand: the proneness of tumours to

appear with advancing life; their predilections for
parts remote from the heart or centre of the nutriment-distributing system; their origin from structures in which the forces of life are running out, as
cicatricial, neoplastic, functionless, and obsolete tissues.
It also coincided that they are often seated at the
ends of bones, and at the ends of divided nerves
where the power of normal growth and development
is deficient; that irritating and exhausting noxæ are
prime causes of tumours. As it marks the end of
developmental life, the breaking up of tissues into
tumour-producing germs is paralleled by reproductive phenomena in plants and micro-organisms,
and by the multiplication of nuclei in the degenerations. The circumstance which makes the ultimate
growth of germs into tumours possible is the relative
feebleness of the germinating tissues. Survival of the
strongest settles the issue of the struggle for nutriment between the old and the young cells.

Though the manifestation of a primordial inherited trait, the division of tissue-cells into germs
must be connected with the influence of the forces
that encompass organisms; heredity decides the *form*
of the dissolution, but circumstances provoke it.

Regarded in this manner, the phenomena at once
fell under the principles of dissolution and evolution.
Tissue-histogenesis is a disintegration, and known in
a multitude of cases to be related to palpable acces-
sions of external energy.

Development and decay in tumours are alternate
processes of evolution and dissolution.

The facts of cystomata admitted of less satisfactory co-ordination. Knowledge of the causes of cysts is in many cases wholly lacking, and some cysts are adaptations rather than diseases. In the frequency with which cysts occur in connection with disused structures, valuable supplementary evidence is furnished of the validity of the hypothesis given above of the origin and nature of tumours. It is more than credible that many cysts originate from the reproductive activity of normal cells, and that the new-born cells undergo liquefaction.

In the remarks upon the teratomata, it was pointed out that if they arise, as there is reason to think, in places where the three germinal layers are in contact, teratomata may be dealt with as true tumours ; but they are initiated during gestation. We also saw that malformations are generally attributable to disorganising forces which act in the intra-uterine life, and that these forms of disease mark a deviation from the norme that corresponds to dissolution.

In the second division of the work, the pathological changes previously viewed reappeared in the guise of special and organic diseases. As harmonising with the interpretation given of the general processes of disease, it was decided that the growth of connective tissue in chronic affections is evolutional, not an inflammatory process. The argument was used that in fibrosis, cirrhosis, and sclerosis of organs and tissues the parenchymatous elements are the first to suffer, and that the interstitial overgrowth is secondary and consequential. Injury or death of the parenchyma favours the growth of interstitial substance simply by diminishing resistance to its growth. Thereupon the data arranged themselves in allied groups, and the organic diseases acquired a new complexion. In the numerous species of kidney and liver disease, facts on every side bore out the interpretation. The distribution of the disease-producing agents, the microscopic morbid morphology, the characters of the interstitial growth and its relations in time to the injury to the parenchyma, supplied a consensus of testimony of considerable weight. And, adding further weight, the supplementary principle to dissolution was seen to be in perfect unison. If accessory specialised structures are more vulnerable than fundamental ones, the death of the parenchyma and survival of the connective-tissue groundwork are according to a requirement of law. No matter whether the changes in question were those of acute nephritis from exanthematous fevers, carbolic acid, or other poisons ; or those of chronic cirrhosis, amyloid disease, and mechanical congestion, the corollary was sustained : the dissolution of the elements was seen to follow the opposite order of their evolution. And in finer detail this was observed of the components of parenchymatous substance. In the kidney the convoluted tubules were seen to be less stable than the straight tubules, and the accessory Malpighian tufts more perishable than the fundamental capillaries.

Bright's disease and hepatic cirrhosis proved to be as dissolution formulates—disintegrations of matter from concomitant absorptions of energy. An essential feature of every form of these affections is the separation of the protoplasm of cells from its former centres of cohesion ; and the facts of ætiology lend a general approval of the proposition that the separation is caused by the reception of surrounding motion. Both in the early and late stages of kidney and liver disease, the normal definiteness, coherence, and heterogeneity of functions and structures is lessened. The growth of interstitial tissue, the various adaptive modifications of function, and the processes of recovery were recognised as the changes of evolution.

Thus it became possible and obligatory to merge all the particular forms of coarse-lesion diseases of the liver and kidney into a general form. The distinction of species which convenience has necessitated was seen to be artificial ; the differences of kind simply mark differences in the conditions of changes that are always the same in type. And the conception of evolution and dissolution as the complements of one another was as lucidly and beautifully verified by the very intricate phenomena of nephritis and hepatitis as by the simple decomposition of a salt in the laboratory. In any attempt to grasp the entirety of the changes in their true natures and mutual relations, this conception cannot indeed be spared.

Thence we proceeded to the consideration of lung-affections, and pneumonia and phthisis were the types selected. The changes in these diseases were observed to be uniform in nature with the changes of nephritis and hepatitis. Reflecting upon the probable causes of pneumonia and phthisis, we apprehended the paths by which noxæ reach the different histological elements of viscera. It was noted that, in the case of the liver, influences that disturb are mostly conveyed to the parenchymatous cells in the blood of the portal vein ; to the parenchymatous cells of the kidney in the blood of the renal arteries. The blood of these vessels, it was noted, is readily affected by the substances taken in as food. The fact was pointed out that *pulmonary* arterial blood is also in intimate intercourse with the ingested aliment, and it was argued that pneumonia and phthisis are connected with the quantity and quality of the blood of the pulmonary artery. While determining and contributory causes were recognised, the opinion was ventured that the necessary internal causes of ordinary pneumonia and phthisis are states of the pulmonary blood as related to states of the pulmonary air, and these states are further to be referred to the environment : to the food eaten and the air respired more especially.

When, in the chapter succeeding, we turned to the diseases of the spinal cord the same general truths were met with. Locomotor ataxia, which we took as representative of a considerable class of neuropathies, was found to correspond to dissolution in the involved changes of function and structure, and probably also in the dependence of these changes on the external agencies with which the organism is in communion. That the destruction of the neural parenchyma in posterior sclerosis is the primary lesion, and the growth of neuroglia a secondary and evolutional phenomenon, is almost beyond such question as might be urged in the case of other fibroid diseases.

Locomotor ataxia and many other affections of the encephalo-spinal system furnished illustrations of the corollary from dissolution; but throughout the entire range of neuropathies the principles of dissolution and evolution were exemplified.

It was not found necessary to make exceptions of the fevers and so-called diathetic disorders—gout, rheumatism, and diabetes. In the case of these, and affections of the mind as well, the principles provided bonds of union with the general and special pathological states that had been dealt with. When at last reference was made to the remaining portions of the field of disease it was declared to be probable that they also might be gathered in.

These, then, are the results. We have been brought to recognise that the multifarious phenomena of disease are originally and innately identical. Neglecting a few aberrant and equivocal examples, true disease is that which dissolution verbally symbolises—a disintegration of matter caused by an absorption of surrounding energy. But in the body as out of it simultaneous and successive with dissolution is the distinct and contrary process of evolution. Together these two processes comprehend not only the changes that are essential to disease, but also its non-essential though inseparable concomitants. What now is the utilitarian outcome of this generalisation? What connection has it with the work of prevention and cure? It is not obviously relevant to the pursuits of those who are practical in this way. Some moot questions of considerable difficulty must be answered before even its indirect bearings upon actual practice can be made clear ; and that it may be ordained as an elementary conception to guide our steps in practice, there will be required a consideration of the processes of disease in fresh aspects. The first matter to be dealt with is the hereditary transmission of disease, and to this we may now turn our attention.

§ 2. HEREDITY AND DISEASE.

Are morbid traits of function and structure passed on from one generation to another as we know healthy traits to be ? Or, to state more exactly the inquiry to be answered, can there be conceded to heredity an equal sphere of action in pathology as in physiology ? And if not, what are the limitations ? These questions are among the most peremptory which modern medicine puts to those who are engaged with its advancement. If the energies of the external world are the true sources of every pathological change—and I think few who have followed me thus far will be disposed to deny this—it is plainly to be seen that a knowledge of the physical factors of disease is of greater or less value as the factor heredity is feeble or active. Diseases dependent upon known external causes are for the most part controllable, but diseases that are inherited are for the most part out of the reach either of measures of prevention or cure. But apparently no one doubts that perhaps the majority of diseases are inherited. Phthisis, gout, rheumatism, diseases of the mind and nervous system, skin diseases, hæmophilia, and others too numerous to mention, are generally believed either to be greatly under the influence of heredity or actually transmitted. It is true that those who are least swayed by belief in morbid inheritance speak of inherited *predispositions* to disease. I hold this to be a weaker form of the same habit of thought, the same mode of interpretation, *minus* the spirit which has been quenched by accumulating experiences. But to the minds of the majority many diseases are as heritable as normal characters. I think it almost certain that heredity plays a very subordinate part in the perpetuation of non-congenital or true disease. The question concerning the transmissibility of morbid states appears to me to have been confused by the failure to distinguish the intrinsic nature and conditions of the dissolutional process.

At the outset of investigation of the subject we are met by the impossibility of entertaining the idea of hereditary transmission in the case of certain common diseases. There is felt an intellectual repulsion towards any suggestion that typhoid fever or acute pneumonia, for example, are due to the operation of heredity. Why is this? Innumerable diseases resembling them in all essentials are every day spoken of as inheritable and inherited. It is because our positive or presumed knowledge of the causes of pneumonia and typhoid fever precludes conception of them as independent of the forces amidst which the organism moves. We are unable to think of these diseases apart from the real or supposed action of the environment. When a disease is known or believed to proceed from sensible external influences heredity cannot be imagined to play any part in its genesis, for that the morbid process may continue, its cause—with which the effects are one and indivisible—must also continue. This cause is not in the microcosm, within which inheritance only works, but in the macrocosm. A still better illustration of the point is afforded by scarlet-fever and rheumatism. The accumulated observations upon the circumstances of its appearance and spread have satisfied physicians that scarlet-fever is in causal relation to something that enters the body from without. Hence no one now thinks of it as inheritable. (I say *now*, for it is not many years since the exanthemata were vaguely looked upon as diseases to which all young flesh is heir.) But rheumatism is not definitely attributable to the environment ; its ætiology is hidden from us, and many, therefore, regard it as handed down by descent. Yet how nearly alike, even in extrinsic characters, these two diseases are, is shown by the existence of scarlatinal rheumatism ; but in intrinsic characters there is no difference between them : they are both disintegrations of the body. Whichever we take of the presumably inherited diseases—cancer, diabetes, epilepsy, chorea, &c.—this we find constantly : it has not yet been associated with the definite action of environing forces ; and whichever we take of the diseases known to depend absolutely upon

the conditions of life we find no genuine belief in
its transmissibility. Speaking generally, for there are
notable exceptions, inheritance is supposed to be most
clearly attested in diseases that are least understood.
As knowledge progresses and true causes come into
the region of light, heredity recedes further and further
from view. Formerly there was prevalent a belief in
the transmissibility of many common ailments which
modern research has shown to arise from ordinary
physical conditions.

Thus we are brought to the conclusion that
heredity is excluded from the ætiology of certain
fevers and inflammations because they are positively
known to proceed from outer actions. But it has
been seen in the course of this work that all inflammations and fevers are disintegrations of the body,
and, with some obscure exceptions, known also to
proceed from outer actions. And almost certainly
the retrograde metamorphoses, the neoplasms, and
organic and functional diseases share this general
character and this condition of inflammations and
fevers. Observe the startling thought called forth
by these reflections. All true diseases may be unin-
heritable, and the constant attribution of non-congenital morbid states to hereditary influences may be
fallacious. How shall we comport ourselves towards
a possible conclusion so momentous and alien to
universal opinion ? Before giving in adherence we
must at least reckon with the reasonable grounds of
that opinion.

Let us first take account of the purest examples
of inherited disease—those which give most counte-

Q

nance to the prevailing belief. These are the cases of racial degeneration in men and animals, of idiot families, of inherited abnormalities like cleft-palate, duplication of fingers, congenital deaf-mutism, &c. Here the deviation from the normal type agrees with dissolution. What is the comment to be offered? That these are congenital, not true diseases. Congenital differs from non-congenital or true disease in that the inferior grade of mental and physical qualities is due, as far as the individual is concerned, to the stage at which development ceases. There is no degradation of the individual life ; the higher type is simply not attained. In true disease, however, there is a loss in the individual of functional and structural traits—such loss as we witness in inflammation, fevers, and diseases generally. These are an unbuilding of the organism ; but in congenital defects and in family and racial degeneration the organism is not unbuilt : it does not build. Inheritance as seen in congenital diseases of th's class cannot, then, be adduced as evidence of the transmissibility of true disease. And of that class of cases represented by congenital syphilis the contention is to be offered that parent and offspring being infected as one organism these are not cases of inheritance.

But the phenomena of many true diseases are thought to support the doctrine of their transmission by descent. Diseases of the vessels, gout, phthisis, paralysis, and others are called to mind. How must be confronted the evidence upon which rests the assent which is given to the supposed inheritance of these ? As the phrase is, gout runs in families, and

so do phthisis, paralysis, and the rest. Here I am
persuaded from the considerations which follow, that
inheritance is simulated in part by the recurrence of
causes. The subtle, unseen factors of gout, phthisis,
insanity, &c., the absorbed energies which set up the
disintegrations of these diseases, are repeated in the
lives of individuals of successive generations, and in
the lives of individuals of collateral family branches,
and there is lent an appearance of hereditary descent.
Of these and very many other affections alleged to be
derived through inheritance it may justly be said
that the causes run in families. This repetition of
pathogenic conditions in the lives of genealogically
related individuals is generally ignored in current
explanations of the origin of diseases by inheritance.
The omission surely vitiates if it does not nullify
every such explanation.

Consistently, it will, I think, be uniformly found
that the general conditions of the lives of those who
suffer from diseases assumed to be inherited *are* the
conditions of the lives of progenitors. The habits
of life of the gouty or phthisical child do not differ
materially from the habits of life of the gouty or
phthisical parent. But we shall learn in the next
section that diseases are repeated in descendants
not solely because the life-conditions are repeated.
Heredity is probably in some degree concerned, but in a
form which it is very important to distinguish from
ordinary inheritance.

To this argument something further may be said,
showing *why* diseases are, presumptively, for the greater
part refractory to transmission by descent. True dis-

eases, we have just seen, cannot be separated from their causes; and causes, being of the environment, are not handed down. But there are additional reasons for the feeble hold which heredity has upon pathological states. When we discriminate between the variations of function and structure that are passed on by parent to offspring and those that are not, we are forced to see that natural selection, working always in confederation with heredity, seizes upon *favourable* variations. Natural selection appropriates organismal acquisitions. But analysis discloses the fact that diseases are losses, not gains; are unfavourable variations, and offer no 'purchase' for the co-operative influence of these two modes of action. Here, then, is another reason for the non-transmissibility of disease. There are others still. In general, diseases unfit the individual for sexual activity by diminishing the surplus energy required for reproduction; therefore the diseased are less prone than the healthy to beget children. Then many diseases are said to be transmitted where transmission is clearly impossible from the circumstances of the cases. Cancer, thought to be heritable, occurs most frequently at the close of sexual life; and pseudo-hypertrophic paralysis, in which hereditary influence is asserted to be well marked, is a disease of childhood—its victims do not propagate. But more important than influences of this sort is that influence which springs from the differences of nature and conditions between normal and abnormal traits. Normal structures were evolved in long periods of time, and have been transmitted through generations unnumbered; therefore, the tendency to their per-

petuation by inheritance must be immensely pre-
dominant over any tendency to the perpetuation by
inheritance of the transitory changes of disease. I
believe that the 'vestiges' of once useful structures
owe their astonishing persistence to the fact that they
have become deeply pressed into the organic arrangement by the selection and transmission of such structures for secular periods. This makes intelligible the
rarity with which deprivation of a limb or other part
leaves any impress upon offspring. Though circum-
cision has been practised among the Jews for ages, it
has not produced congenital preputial imperfection in
the race. Nor do we ever find that amputation of a
limb or loss of the cortex of the kidney from Bright's
disease is followed by corresponding anatomical deficiencies in children. In the African transported
to northern latitudes, the dark skin persists through
indeterminable generations—provided there is no
cross-breeding—but the endemic diseases of his race
are not transported with him.

Hitherto all reasoning upon the heritableness of
diseases has proceeded on the tacit assumption that
morbid changes are subject to the same law of vital
action as healthy changes. It has been discovered,
however, that the two are dissimilar both in nature
and in the circumstances of their genesis. The
traits we every day recognise as inherited are the
results of an infinity of co-ordinate actions. There
may be instanced the bony framework of the face,
the colour of the iris, the gait, special mental apti-
tudes. All these, and attributes of the same order,
represent a vast integration of forces, groups of

organised energies. It is this organisation which gives them individuality and makes their hereditary transmission possible. They are, in other words, self-existent, have being independent of the original conditions out of which they grew. Thus, that inherited functions and structures may unfold themselves in the developing embryo the only pre-requisites are certain general conditions of blood-supply, temperature, &c. The special conditions of their evolution are not required. Vision is possible to the human infant immediately after birth, the eyes being developed *in utero* without the aid of that light-energy which was all essential for their evolution in the race. Now it is quite contrary with most true diseases conceived as processes of dissolution. These are disorganisations, and are therefore not self-existent. To what extent tumours as true diseases are possible exceptions may be left for future consideration.

So that many direct and indirect pieces of evidence cohere to defend the inference, necessitated by the nature and conditions of disease, that heredity is probably very feebly concerned in the continuance of morbid states. The universal belief in the heritableness of disease is probably not a faithful expression of the truth. We may, then, for the time being and the purposes of future work, conclude that diseases are the issue of interactions between the individual and his environment ; that even in the individual, ' the deepest causes of true diseases lie external to the organism.' Of diseases in the race it has been largely shown inductively in previous chapters that they are in the first instance externally derived ; and this fact requires us to remember that heredity where proved to account for disease in the individual is never a *vera causa*. It is the transmission of attributes from one generation to another, but is not itself originative; we are in such a case referred back to the lives of individual ancestors in the search for causes.

Out of this argument the reader has doubtless culled the principle of practice it contains. If true diseases are rarely inherited and they are the proceeds of actions and reactions between the body and forces that fall upon it, the ruling precept of practical medicine must clearly be the detection and removal of causes. This is the fundamental principle we shall presently enforce ; but before doing so there must be considered one which is ancillary to it in the hands of the practitioner—the principle of organic equilibration, or *vis medicatrix naturæ*.

§ 3. ORGANIC EQUILIBRATION.

We have observed again and again that dissolution and evolution are co-existent and sequent to one another. As the various changes of disease passed before us and we distinguished the two orders into which they may be separated, we were led to regard the growth of connective tissue in the organic affections as an adaptive change. Not of malign influence, but subserving the welfare of the organism. We placed it in the same category as resolution and repair. And a like disposition was made of those processes by which functional equilibrium is restored in the acute systemic diseases. Now these several kinds of action are, from the practical point of view, best conceived as illustrations of the principle of equilibration. The principle of equilibration in its most developed form we owe to Mr. Spencer; and it will, I think, be seen to lie at the very root of sound medical conceptions, and to give a meaning of singular beauty to many of the common phenomena of disease.

Equilibration is exhibited throughout all nature; and repair, resolution, adaptation, &c., are examples of it upon the view that life consists in the maintenance of an adjustment between inner functions and outer incident forces. For illustrations from the inorganic world the reader may turn to 'First Principles,' chap. xxii. The following extract is from Mr. Spencer's exposition of equilibration as we witness it in the living body: 'The sensible motion constituting each visible action of an organism is soon brought to a close by some adverse force within or without the organism. When the arm is raised, the motion given to it is antagonised partly by gravity and partly by the internal resistances consequent on structure; and its motion, thus suffering continual deduction, ends when the arm has reached a position at which the forces are equilibrated. The limits of each systole and diastole of the heart severally show us a momentary equilibrium between muscular strains that produce opposite movements; and each gush of blood requires to be immediately followed by another, because the rapid dissipation of its momentum would otherwise soon bring the mass of circulating fluid to a stand. As much in actions and reactions going on among the internal organs as in the mechanical balancing of the whole body, there is at every instant a progressive equilibration of the motions at every instant produced. Viewed in their aggregate, and as forming a series, the organic functions constitute a dependent moving equilibrium—a moving equilibrium of which the motive power is ever being dissipated through the special equilibrations just exemplified, and is ever being renewed by the taking-in of additional motive power. Food is a store of force which continually adds to the momentum of the vital actions as much as is deducted from them by the forces overcome. All the functional movements thus maintained are, as we have seen, rhythmical (§ 85); by their union compound rhythms of various lengths and complexities are produced, and in these simple and compound rhythms the process of equilibration, besides being exemplified at each extreme of every rhythm, is seen in the habitual preservation of a constant mean, and in the re-establishment of that mean when accidental causes have produced divergence from it. When, for instance, there is a great expenditure of motion through muscular activity, there arises a reactive demand on those stores of latent motion which are laid up in the form of consumable matter throughout the tissues; increased respiration and increased rapidity of circulation are instrumental to an extra genesis of force that counterbalances the extra dissipation of force. This unusual transformation of molecular motion into sensible motion is presently followed by an unusual absorption of food—the source of molecular motion ; and in proportion as there has been a prolonged draft upon the spare capital of the system is there a tendency to a prolonged rest during which that spare capital is replaced. If the deviation from the ordinary course of the functions has been so great as to derange them, as when violent exertion produces loss of appetite and loss of sleep, an equilibration is still eventually effected. Providing the disturbance is not such as to overturn the balance of the functions, and destroy life (in which case a complete equilibration is suddenly effected), the ordinary balance is by-and-by re-established ; the returning appetite is keen in proportion as the waste has been large ; while sleep, sound and prolonged, makes up for previous wakefulness. Not even in those extreme cases where some excess has wrought a derangement that is never wholly rectified is there an exception to the general law ; for in such cases the cycle of the functions is after a time equilibrated about a new mean state, which thenceforth becomes the normal state of the individual. Thus, among the involved rhythmical changes constituting organic life, any disturbing force that works an excess of change in some direction is gradually diminished, and finally neutralised by antagonistic forces, which thereupon work a compensating change in the opposite direction, and so, after more or less of oscillation, restore the medium condition. And this process it is which constitutes what physicians call the *vis medicatrix naturæ*.'

Now the common formative changes of disease are plain manifestations of this principle of organic action.

When the parenchymatous substance of an organ is destroyed, there is a disturbance of the statical balance of forces, and the interstitial tissue may grow out to restore the equilibrium. And when the heart enlarges in arterial and valvular disease, there is an analogous structural adjustment to re-establish the equilibrium of the circulation. But the principle obtains *whenever the normal play of the functions is disordered.* The state of health is one to which the energies of the system are bent with the highest strength of the organism. It seems inevitable that since the type of physiological action we call the norme is the product of all ancestral intercourse with the environment, this type of action must present considerable stability in the face of forces tending to change it. As these forces are insignificant in comparison with the aggregate of forces which have produced normal functions, they cannot, unless they annihilate them, do more than temporarily modify the moving equilibrium. (This is in harmony with the argument on the inheritance of pathological states, that ordinarily the changes of disease in parents are obliterated in offspring by the overwhelming organisation of evolutionally produced actions.) A poisonous but not fatal dose of alcohol sets up a new series of rhythms in the functional systems. Presently the poison is consumed or expelled, and there is a gradual re-establishment of the previous balance. So, too, of the effects of the agents that cause acute disease. When, in chronic disease, the perturbations are continuous, and do not greatly depart from the mean of normal physiological actions, there is also a more or less successful

concert of predominating energies to preserve that
mean.

Still, it may be said, granting the existence of
self-restorative powers such as here described in
general form—they are vaguely recognised by most
people—we do not know of them in detail. Yet the
individual actions by means of which the general
balance of functions is recovered—as in the specific
fevers—must be physical actions, assimilable with
familiar physiological changes. They are so, undoubtedly ; and I believe the most important of them
are the common forms of physiological perturbation.
We may look at some concrete instances.

The pyrexia of fever, when brought into relation
with its causes and the needs of the organism to deal
with those causes, may be construed as conservative
and beneficent. It is, I think, in the highest degree
probable that febrile heat, whatever the mechanism of
its production, facilitates the redistribution of matter
and motion involved in equilibrating processes. Take
also the sweating of fever—of ague and acute rheumatism. This is clearly to be interpreted as a
consequence of the presence in the blood of matter
inimical to normal function, and, as an adjustment of
function, is calculated to assist in the elimination
of the deleterious substances. The sweating is only
relatively baneful—that is, when contrasted with
the functions of the skin in health ; relatively to its
cause, it is desirable. It is astonishing how many of
the daily occurring forms of disease may likewise be
conceived as forms of equilibration. Connected with
its antecedents, suppuration is to be understood as a

serviceable process. If it arises from the action of micro-organisms or poisons, it is probably useful in localising the distribution of such noxæ—that is, prevents a systemic distribution which would perhaps be fatal. This is the point of view assumed in the recently advanced idea that the leucocytes of pus are 'phago-cytes.' The granuloma also is a species of reaction on the part of the tissues to locally incident noxæ, and may be, in reference to the latter, like suppuration, the lesser of two evils. That inflammation, unmistakably disintegrative and dissolutional, is relatively good appears to be shown by the fact that from the milder forms of it there may be perfect recovery. Admiration should be the uppermost feeling as we contemplate the fertility and delicacy of adjustment implied by resolution. Dropsy is certainly a functional adaptation. Without the determination of fluids to the connective-tissue spaces death would seem to be inevitable under the physiological circumstances of dropsy. As the effects of their proximate conditions, a multitude of symptoms and diseases are favourable reactions. So ought we to regard the hæmatemesis of hepatitis, the epistaxis of puberty, the hæmoptysis of phthisis and heart-disease. Cough and pulmonary expectoration, the discharge of matter through the kidneys and bowels in the fevers and chronic diseases, are all relatively advantageous— advantageous when joined with their causes. Illustrations might be taken from almost every disease, but these will have served to establish the point of view. A qualification must, however, be inserted if we are to justly estimate abnormal processes as thus

conceived. Wonderful as is the consentancity of the
actions involved, organic equilibration is neither in-
telligent nor purposive. The best examples of it in
disease are simply the least unfavourable among the
pathological changes. And the most deeply organised
of them, those that have presumably been taken hold
of by selective action and heredity (see *infra*), are
plastic and readily perverted. Only under certain conditions and without design does the disordered body
make for health. The presence of apparently insignificant and trivial circumstances will oftentimes give
fatal misdirection to the forces of the organism.

And now, with this conception before the mind, it
is possible to fulfil the promise of the previous section
—that is, to say in what manner heredity is probably
most concerned in the recurrence of diseases in
descendants. It is, I believe, in this way. *To similar causes there are in those similarly constituted*—like
members of families—*similar modes of physiologial or
pathological reaction.* The familiar types of equilibration, as the emesis of poisoning, the high temperature
of fever, the dropsy of heart-disease and nephritis, &c.,
are perhaps inherited. I have elsewhere suggested
that natural selection would share in the work of establishing reparative processes in wounds. ‘Animals
whose wounds were not repaired would strive in the
struggle for existence under a heavy disability. The
non-union of muscles torn and limbs broken during
encounters with prey and enemies would lead to
incapacity for further encounters, and a dying off
without issue; while others, whose injuries were
repaired, would escape the direct hindrances and

indirect dangers to life entailed by unhealed wounds, and would reproduce their kind.' In the same place it is also pointed out that selective and hereditary actions have probably been connected not only with the organisation of modes of adjustment in common diseases, like scarlet fever and measles, but also with the establishing of personal insusceptibilities to the influence of their causes. And I think it may well be supposed that natural selection and heredity have conspired to aid in the survival of such generic forms of equilibration as those we have just now noticed. Given in the lives of parent and child the necessary external causes of gout, or phthisis, or cancer, and there will be approximately the same forms of reaction to them, the same diseases in members of the same family. No one will confound this with the common idea of the inheritance of disease ; nor should it, it appears to me, be confounded with the idea of inherited predisposition to disease. As already argued (§ 2), diseases for the most part are not self-existent, and their reappearance in families is determined almost wholly by recurrent environmental conditions. The predispositions of the organism are to normality, not to abnormality. In this view, diseases are less evil than the causes of them, and are presented to us as signifying a contest between the powers within and the powers without.

§ 4. PRACTICAL DEDUCTIONS AND CONCLUSION.

What of instant utility may we learn from this research ? Can anything be deduced that will promise to give greater consistency and certitude to

medical practice ? We seek for the highest conclusions of philosophy in conclusions to act. That
disease is the expression of a relation between the
individual and his surroundings, and that in all
pathogenesis there have been external factors, adequate proof has, I think, come before us. Must we
not, then, both for prevention and cure, ever look at
disease from the standpoint of this relation ? There
appears to be no escape from the deduction that,
given a knowledge of the circumstances of their
genesis, a very large majority of diseases might be
avoided. And the positive achievements of modern
hygiene ratify this deduction. It is frequently
possible to so change the organism's conditions,
while fulfilling the conditions of health, as to insure
prophylaxis. By the same provision, probably, most
acquired diseases, prior to the stage of permanent
loss of structure, are curable. *Cessante causa, cessat
effectus*. This is true for every symptom and for
every group of symptoms. In each case the quest
then should be for causes. Often diseases are
ætiologically related to the conditions of life only in
a remote sense. Such is sometimes the case with
calculus, tumours, effusions into cavities, &c. Here
we seek out and remove the anatomical causes. But
these products of morbid action are truly the effects
of an imperfect adjustment to environment, and
many of them are preventable by the best uses of
existing knowledge that recognises the dependence
of disease upon external agencies. Therefore the
higher principle is that which constrains us to fix
our eyes on morbid phenomena as parts of an

unbroken series of changes connected indissolubly with the energies of the outer world.

The present inquiry teaches that diseases are not self-originated and self-maintained, and that the powers of the physician are therefore not limited by the inexorableness of disease. The first condition for dealing effectually with many common scourges and crippling ailments would be to grasp firmly the necessary circumstances of their origin and being. Fortunately, we are invited to researches in pathogenesis by the relative simplicity of the process of dissolution. Though the factors of disease are often inexplicable to the most diligent and skilled investigators, yet when we compare the conditions out of which evolutions and dissolutions respectively arise, those of the former are perceived to be vastly the more complex. In everyday parlance, destruction is more simple than construction. Look at the many interconnected actions involved in the formation of some ordinary evolutional product—say, the arrangement of the type of this page. The actions are of varied kinds, and duly and reciprocally proportioned. But how simple may be the dissolution of the structure. A single form of energy, as fire, or mechanical motion, will undo in the briefest time the work of the proportioned actions And the destroying force is without definite relations to the forces that oppose it. It is the same in the organic world. The kinds and consensus of energies required for the production of any bodily structure— for example, a renal organ — are wholly beyond computation ; but a measurable quantity of carbolic

acid will loosen all its combined activities and disintegrate its substance; will produce the disease we
call nephritis. Throughout we find a similar relative simplicity in the causation of diseases. Syphilis
depends on the inception of a virus; paralysis follows
the extravasation of blood from a ruptured vessel;
insanity, the shock of a bereavement. Absolutely,
however, diseases are often most complex in causation, and ætiological investigation is beset with difficulties. There is not infrequently a multiplicity of
causes, and intricate sequences of them. And life is
a *moving* equilibrium. Disease, as the physician encounters it, is always dynamic; hence in part the
variableness of its phenomena. The physiological
term of the relation between inner functions and
outer forces eludes the observer by its mobility. An
equal difficulty lies in the complicated relations of
the internal changes of disease, in the order of
succession of the changes, and in the nature of the
external causes. A great number of those diseases,
whose ætiology is as yet closed to us, are undoubtedly the consequences of extremely unobtrusive habits and circumstances. Influences coming
from without, and suspected by only a few observers,
are, I think, the efficient instigators and maintain-
ers of unhealthy processes that maim and destroy
thousands. Chief among the as yet unrecognised
influences I should place those in action upon the
fundamental functions of alimentation, oxidation, and
excretion. In everything environment sums up—the
food we eat, the air we breathe, the work we do, &c.
—and in the mutual interactions of the involved

functions, there may be discovered the means of prevention and cure in a host of maladies, bodily and mental. It is an apothegm of Dr. Hughlings Jackson that diseases are flaws in an evolutionary system. Within the known boundaries of this system there is, I believe, a body of knowledge with vast and momentous implications for the physician.

From all that has passed before us, it must be admitted that the thing of most importance in practical medicine, equally for therapeusis · as for prophylaxis, is to know the organismal and environmental conditions out of which diseases are born. Not the internal factors alone, or the external factors alone, but the two sets of factors in relation. Yet hardly of less consequence in the management of cases must be a scientific conception and utilisation of the self-restorative powers. It is anomalous and a reproach that only the scantiest allusions to the processes of organic equilibration are to be found in text-books. No place is given to the general principle. Yet without an understanding of it in its general and special applications all thought about diseases in their dynamical aspects must of necessity be rude and erring. It is not hazardous to say that most of the empiricism of medicine is traceable to ignorance of the living body as a conservative arrangement of forces. Homœopathy, allopathy, hydropathy, and kindred dogmatic systems will continue as exclusive methods of practice as long as the *vis medicatrix naturæ* remains an unrecognised principle of medical science.

While insisting upon this conservatism of the

organism as an agency in the cure of morbid states
it will be well to revert to some remarks made
towards the close of the last section. Serious will
be the misapplication of these considerations if they
are deemed to approve what is known as the 'ex-
pectant' method of treatment. This would be to
assume perfection in natural processes that are the
outcome of natural weakness. Rightly, the unintel-
ligent though bounteous operations of the *vis medicatrix naturæ* must be under the empire of the intel-
ligent mind. Purposeless, blind, and with a narrow
range of action, this organised power, while entering
into all our calculations in treating disease, should
be the mere instrument of the physician, controlled
and directed by him. It must be subsidiary to his
intelligence—an intelligence which can lead him to
conclusions utterly beyond its reach. Organic equilibration unaided by knowledge and skill may be
powerless to prevent the formation of a calculus
or to remove such a body when impacted in the
pelvis of the kidney ; and we have an example of the
limitations of its beneficence where the life-conserv-
ing cardiac hypertrophy of Bright's disease assists
in the fatal bursting of a cerebral blood-vessel.

It will not pass unobserved that in pathology
organic equilibration also requires the important
practical view that disease marks an inability of the
organism to cope with the energies that invest it.
Disease is a struggle between the body and its conditions ; hence the conditions are weakened as the
organism is strengthened. Here is recognised another internal factor in the cure of disease. Though

external causes demand, as we have given to them, paramount attention, the *relativity* of the disease-producing influence of environing forces has an important and many-sided bearing on treatment. Agencies that are maleficent under one set of circumstances may be harmless or beneficent under another. The estimation of these variables would appear to be in the special province of practical medicine. The physiological norme itself is variable in historical time and geographical space and in the lives of individuals ; and since it is the sum of co-ordinate conditions, the removal of the causes of disease is the institution of determinately proportioned circumstances, which, theoretically, should never be exactly repeated in the cases treated. Here, surely, may be found an unanswerable argument against therapeutic rules and dogmas. We are not led to believe that there can be an enduring system of therapeutics based on the physiological influence of chemical and physical forces in artificial combinations, precious as are some of the rational and empiric truths the system embodies. It is only required that we should understand pathology to see that diseases and symptoms are not separate and individual, to be treated separately and individually, but that they are phenomena constituting a series, each member of which will persist as long as its antecedents persist. Whether the restoration to health is compassed by natural equilibration or by art, knowledge of the processes involved should in either case be received into the institutes of medicine.

If we have concluded truthfully, the most necessary knowledge for the rational treatment of disease is physiology and pathology as one science in its widest and deepest connections—wide, as including the phenomena and interrelations of health and disease ; and deep, as drawing the sap of its life from biology, chemistry, and physics. Concerning those useful means of medical treatment that do not obey the canons enunciated, they would have place in such a scheme where a rational justification is possible by appeals to settled truths ; but empiricism, serviceable as well as hurtful, has a position outside the precincts of the estate of science.

PRINTED BY
SPOTTISWOODE AND CO., NEW-STREET SQUARE
LONDON

Catalogue of Books

PUBLISHED BY

MESSRS. LONGMANS, GREEN, & CO.

39 PATERNOSTER ROW, LONDON, E.C.

Abbey.—*THE ENGLISH CHURCH AND ITS BISHOPS,* 1700-1800. By CHARLES J. ABBEY, Rector of Checkendon. 2 vols. 8vo. 24*s.*

Abbey and Overton.—*THE ENGLISH CHURCH IN THE EIGHTEENTH CENTURY.* By CHARLES J. ABBEY, Rector of Checkendon, and JOHN H. OVERTON, Rector of Epworth and Canon of Lincoln. Crown 8vo. 7*s.* 6*d.*

Abbott.—*THE ELEMENTS OF LOGIC.* By T. K. ABBOTT, B.D. 12mo. 3*s.*

Acton. — *MODERN COOKERY FOR PRIVATE FAMILIES.* By ELIZA ACTON. With 150 Woodcuts. Fcp. 8vo. 4*s.* 6*d.*

A. K. H. B.—*THE ESSAYS AND CONTRIBUTIONS OF A. K. H. B.*—Uniform Cabinet Editions in crown 8vo.
Autumn Holidays of a Country Parson, 3*s.* 6*d.*
Changed Aspects of Unchanged Truths, 3*s.* 6*d.*
Commonplace Philosopher, 3*s.* 6*d.*
Counsel and Comfort from a City Pulpit, 3*s.* 6*d.*
Critical Essays of a Country Parson, 3*s.* 6*d.*
Graver Thoughts of a Country Parson. Three Series, 3*s.* 6*d.* each.
Landscapes, Churches, and Moralities, 3*s.* 6*d.*
Leisure Hours in Town, 3*s.* 6*d.*
Lessons of Middle Age, 3*s.* 6*d.*
Our Little Life. Two Series, 3*s.* 6*d.* each.
Our Homely Comedy and Tragedy, 3*s.* 6*d.*
Present Day Thoughts, 3*s.* 6*d.*
Recreations of a Country Parson. Three Series, 3*s.* 6*d.* each.
Seaside Musings, 3*s.* 6*d.*
Sunday Afternoons in the Parish Church of a Scottish University City, 3*s.* 6*d.*

Amos.—*WORKS BY SHELDON AMOS.*

A PRIMER OF THE ENGLISH CONSTITUTION AND GOVERNMENT. Crown 8vo. 6*s.*

A SYSTEMATIC VIEW OF THE SCIENCE OF JURISPRUDENCE. 8vo. 18*s.*

Aristotle.—*THE WORKS OF.*

THE POLITICS, G. Bekker's Greek Text of Books I. III. IV. (VII.) with an English Translation by W. E. BOLLAND, M.A. ; and short Introductory Essays by A. LANG, M.A. Crown 8vo. 7*s.* 6*d.*

THE POLITICS ; Introductory Essays. By ANDREW LANG. (From Bolland and Lang's ' Politics.') Crown 8vo. 2*s.* 6*d.*

THE ETHICS ; Greek Text, illustrated with Essays and Notes. By Sir ALEXANDER GRANT, Bart. M.A. LL.D. 2 vols. 8vo. 32*s.*

THE NICOMACHEAN ETHICS, Newly Translated into English. By ROBERT WILLIAMS, Barrister-at-Law. Crown 8vo. 7*s.* 6*d.*

Armstrong.—*WORKS BY GEORGE FRANCIS ARMSTRONG, M.A.*

POEMS : Lyrical and Dramatic. Fcp. 8vo. 6*s.*

KING SAUL. (The Tragedy of Israel, Part I.) Fcp. 8vo. 5*s.*

KING DAVID. (The Tragedy of Israel, Part II.) Fcp. 8vo. 6*s.*

KING SOLOMON. (The Tragedy of Israel, Part III.) Fcp. 8vo. 6*s.*

UGONE : A Tragedy. Fcp. 8vo. 6*s.*

A GARLAND FROM GREECE ; Poems. Fcp. 8vo. 9*s.*

STORIES OF WICKLOW ; Poems. Fcp. 8vo. 9*s.*

VICTORIA REGINA ET IMPERATRIX: a Jubilee Song from Ireland, 1887. 4to. 5*s.* cloth gilt.

THE LIFE AND LETTERS OF EDMUND J. ARMSTRONG. Fcp. 8vo. 7*s.* 6*d.*

Armstrong.—*WORKS BY EDMUND J. ARMSTRONG.*

POETICAL WORKS. Fcp. 8vo. 5*s.*

ESSAYS AND SKETCHES. Fcp. 8vo. 5*s.*

A

Arnold. — *WORKS BY THOMAS ARNOLD, D.D. Late Head-master of Rugby School.*

INTRODUCTORY LECTURES ON MODERN HISTORY, delivered in 1841 and 1842. 8vo. 7s. 6d.

SERMONS PREACHED MOSTLY IN THE CHAPEL OF RUGBY SCHOOL. 6 vols. crown 8vo. 30s. or separately, 5s. each.

MISCELLANEOUS WORKS. 8vo. 7s. 6d.

Arnold.—*A MANUAL OF ENGLISH LITERATURE,* Historical and Critical. By THOMAS ARNOLD, M.A. Crown 8vo. 7s. 6d.

Arnott.—*THE ELEMENTS OF PHYSICS OR NATURAL PHILOSOPHY.* By NEIL ARNOTT, M.D. Edited by A. BAIN, LL.D. and A. S. TAYLOR, M.D. F.R.S. Woodcuts. Crown 8vo. 12s. 6d.

Ashby. — *NOTES ON PHYSIOLOGY FOR THE USE OF STUDENTS PREPARING FOR EXAMINATION.* With 120 Woodcuts. By HENRY ASHBY, M.D. Lond. Fcp. 8vo. 5s.

Atelier (The) du Lys; or, an Art Student in the Reign of Terror. By the Author of ' Mademoiselle Mori.' Crown 8vo. 2s. 6d.

Bacon.—*THE WORKS AND LIFE OF.*

COMPLETE WORKS. Edited by R. L. ELLIS, M.A. J. SPEDDING, M.A. and D. D. HEATH. 7 vols. 8vo. £3. 13s. 6d.

LETTERS AND LIFE, INCLUDING ALL HIS OCCASIONAL WORKS. Edited by J. SPEDDING. 7 vols. 8vo. £4. 4s.

THE ESSAYS; with Annotations. By RICHARD WHATELY, D.D., 8vo. 10s. 6d.

THE ESSAYS; with Introduction, Notes, and Index. By E. A. ABBOTT, D.D. 2 vols. fcp. 8vo. price 6s. Text and Index only, without Introduction and Notes, in 1 vol. fcp. 8vo. 2s. 6d.

Bagehot. — *WORKS BY WALTER BAGEHOT, M.A.*

BIOGRAPHICAL STUDIES. 8vo. 12s.

ECONOMIC STUDIES. 8vo. 10s. 6d.

LITERARY STUDIES. 2 vols. 8vo. 28s.

THE POSTULATES OF ENGLISH POLITICAL ECONOMY. Crown 8vo. 2s. 6d.

Bagwell. — *IRELAND UNDER THE TUDORS,* with a Succinct Account of the Earlier History. By RICHARD BAGWELL, M.A. Vols. I. and II. From the first invasion of the Northmen to the year 1578. 2 vols. 8vo. 32s.

The BADMINTON LIBRARY, edited by the DUKE OF BEAUFORT, K.G. assisted by ALFRED E. T. WATSON.

Hunting. By the DUKE OF BEAUFORT, K.G. and MOWBRAY MORRIS. With Contributions by the Earl of Suffolk and Berkshire, Rev. E. W. L. Davies, Digby Collins, and Alfred E. T. Watson. With Coloured Frontispiece and 53 Illustrations by J. Sturgess, J. Charlton, and Agnes M. Biddulph. Crown 8vo. 10s. 6d.

Fishing. By H. CHOLMONDELEY-PENNELL. With Contributions by the Marquis of Exeter, Henry R. Francis, M.A., Major John P. Traherne, G. Christopher Davies, R. B. Marston, &c.

Vol. I. Salmon, Trout, and Grayling. With 150 Illustrations. Cr. 8vo. 10s. 6d.

Vol. II. Pike and other Coarse Fish. With 58 Illustrations. Cr. 8vo. 10s. 6d.

Racing and Steeplechasing. By the EARL OF SUFFOLK, W. G. CRAVEN, The Hon. F. LAWLEY, A. COVENTRY, and A. E. T. WATSON. With Coloured Frontispiece and 56 Illustrations by J. Sturgess. Cr. 8vo. 10s. 6d.

Shooting. By Lord WALSINGHAM and Sir RALPH PAYNE - GALLWEY, with Contributions by Lord Lovat, Lord Charles Lennox Kerr, The Hon. G. Lascelles, and Archibald Stuart Wortley. With 21 full-page Illustrations and 149 Woodcuts by A. J. Stuart-Wortley, C. Whymper, J. G. Millais, &c.

Vol. I. Field and Covert. Cr. 8vo. 10s. 6d.

Vol. II. Moor and Marsh. Cr. 8vo. 10s. 6d.

Cycling. By VISCOUNT BURY, K.C.M.G. and G. LACY HILLIER. With 19 Plates and 61 Woodcuts by Viscount Bury and Joseph Pennell. Cr. 8vo. 10s. 6d.

Athletics and Football. By MONTAGUE SHEARMAN. With Introduction by Sir Richard Webster, Q.C. M.P. With 6 full-page Illustrations and 45 Woodcuts from Drawings by Stanley Berkeley, and from Instantaneous Photographs by G. Mitchell. Cr. 8vo. 10s. 6d.

*** Other volumes in preparation.

Bain. — *WORKS BY ALEXANDER BAIN, LL.D.*

MENTAL AND MORAL SCIENCE; a Compendium of Psychology and Ethics. Crown 8vo. 10s. 6d.

THE SENSES AND THE INTELLECT. 8vo. 15s.

[Continued on next page.

Bain. — *WORKS BY ALEXANDER BAIN, LL.D.—continued.*

THE EMOTIONS AND THE WILL. 8vo. 15*s*.

PRACTICAL ESSAYS. Cr. 8vo. 4*s*. 6*d*.

LOGIC, DEDUCTIVE AND INDUCTIVE. PART I. *Deduction*, 4*s*. PART II. *Induction*, 6*s*. 6*d*.

JAMES MILL; a Biography. Cr. 8vo. 2*s*.

JOHN STUART MILL; a Criticism, with Personal Recollections. Cr. 8vo. 1*s*.

Baker. — *WORKS BY SIR SAMUEL W. BAKER, M.A.*

EIGHT YEARS IN CEYLON. Crown 8vo. Woodcuts. 5*s*.

THE RIFLE AND THE HOUND IN CEYLON. Crown 8vo. Woodcuts. 5*s*.

Bale. — *A HANDBOOK FOR STEAM USERS;* being Notes on Steam Engine and Boiler Management and Steam Boiler Explosions. By M. POWIS BALE, M.I.M.E. A.M.I.C.E. Fcp. 8vo. 2*s*. 6*d*.

Ball. — *THE REFORMED CHURCH OF IRELAND* (1537–1886). By the Right Hon. J. T. BALL, LL.D. D.C.L. 8vo. 7*s*. 6*d*.

Barker. — *A SHORT MANUAL OF SURGICAL OPERATIONS*, having Special Reference to many of the Newer Procedures. By ARTHUR E. J. BARKER, F.R.C.S. Surgeon to University College Hospital. With 61 Woodcuts in the Text. Crown 8vo. 12*s*. 6*d*.

Barrett. — *ENGLISH GLEES AND PART-SONGS.* An Inquiry into their Historical Development. By WILLIAM ALEXANDER BARRETT. 8vo. 7*s*. 6*d*.

Beaconsfield. — *WORKS BY THE EARL OF BEACONSFIELD, K.G.*

NOVELS AND TALES. The Hughenden Edition. With 2 Portraits and 11 Vignettes. 11 vols. Crown 8vo. 42*s*.

Endymion.
Lothair.
Coningsby.
Sybil.
Tancred.
Venetia.

Henrietta Temple.
Contarini Fleming, &c.
Alroy, Ixion, &c.
The Young Duke, &c.
Vivian Grey.

NOVELS AND TALES. Cheap Edition, complete in 11 vols. Crown 8vo. 1*s*. each, boards; 1*s*. 6*d*. each, cloth.

THE WIT AND WISDOM OF THE EARL OF BEACONSFIELD. Crown 8vo. 1*s*. boards, 1*s*. 6*d*. cloth.

Becker. — *WORKS BY PROFESSOR BECKER, translated from the German by the Rev. F. METCALF.*

GALLUS; or, Roman Scenes in the Time of Augustus. Post 8vo. 7*s*. 6*d*.

CHARICLES; or, Illustrations of the Private Life of the Ancient Greeks. Post 8vo. 7*s*. 6*d*.

Bentley. — *A TEXT-BOOK OF ORGANIC MATERIA MEDICA.* Comprising a Description of the VEGETABLE and ANIMAL DRUGS of the BRITISH PHARMACOPŒIA, with some others in common use. Arranged Systematically and especially Designed for Students. By ROBT. BENTLEY, M.R.C.S.Eng. F.L.S. With 62 Illustrations. Crown 8vo. 7*s*. 6*d*.

Boultbee. — *A COMMENTARY ON THE 39 ARTICLES* of the Church of England. By the Rev. T. P. BOULTBEE, LL.D. Crown 8vo. 6*s*.

Bourne. — *WORKS BY JOHN BOURNE, C.E.*

CATECHISM OF THE STEAM ENGINE in its various Applications in the Arts, to which is now added a chapter on Air and Gas Engines, and another devoted to Useful Rules, Tables, and Memoranda. Illustrated by 212 Woodcuts. Crown 8vo. 7*s*. 6*d*.

HANDBOOK OF THE STEAM ENGINE; a Key to the Author's Catechism of the Steam Engine. With 67 Woodcuts. Fcp. 8vo. 9*s*.

RECENT IMPROVEMENTS IN THE STEAM ENGINE. With 124 Woodcuts. Fcp. 8vo. 6*s*.

Bowen. — *HARROW SONGS AND OTHER VERSES.* By EDWARD E. BOWEN. Fcp. 8vo. 2*s*. 6*d*.; or printed on hand-made paper, 5*s*.

Brabazon. — *SOCIAL ARROWS:* Reprinted Articles on various Social Subjects. By Lord BRABAZON. Crown 8vo. 1*s*. boards, 5*s*. cloth.

Brassey. — *WORKS BY LADY BRASSEY.*

A VOYAGE IN THE 'SUNBEAM,' OUR HOME ON THE OCEAN FOR ELEVEN MONTHS.

Library Edition. With 8 Maps and Charts, and 118 Illustrations, 8vo. 21*s*.

Cabinet Edition. With Map and 66 Illustrations, crown 8vo. 7*s*. 6*d*.

School Edition. With 37 Illustrations, fcp. 2*s*. cloth, or 3*s*. white parchment with gilt edges.

Popular Edition. With 60 Illustrations, 4to. 6*d*. sewed, 1*s*. cloth.

[*Continued on next page.*

Brassey. — *WORKS BY LADY BRASSEY*—continued.

SUNSHINE AND STORM IN THE EAST.
Library Edition. With 2 Maps and 114 Illustrations, 8vo. 21s.
Cabinet Edition. With 2 Maps and 114 Illustrations, crown 8vo. 7s. 6d.
Popular Edition. With 103 Illustrations, 4to. 6d. sewed, 1s. cloth.

IN THE TRADES, THE TROPICS, AND THE 'ROARING FORTIES.'
Library Edition. With 8 Maps and Charts and 292 Illustrations, 8vo. 21s.
Cabinet Edition. With Map and 220 Illustrations, crown 8vo. 7s. 6d.
Popular Edition. With 183 Illustrations, 4to. 6d. sewed, 1s. cloth.

THREE VOYAGES IN THE 'SUNBEAM.'
Popular Edition. With 346 Illustrations, 4to. 2s. 6d.

Browne.—*AN EXPOSITION OF THE 39 ARTICLES*, Historical and Doctrinal. By E. H. BROWNE, D.D., Bishop of Winchester. 8vo. 16s.

Bryant.—*EDUCATIONAL ENDS;* or, the Ideal of Personal Development. By SOPHIE BRYANT, D.Sc.Lond. Crown 8vo. 6s.

Buckle. — *WORKS BY HENRY THOMAS BUCKLE.*

HISTORY OF CIVILISATION IN ENGLAND AND FRANCE, SPAIN AND SCOTLAND. 3 vols. crown 8vo. 24s.

MISCELLANEOUS AND POSTHUMOUS WORKS. A New and Abridged Edition. Edited by GRANT ALLEN. 2 vols. crown 8vo. 21s.

Buckton.—*WORKS BY MRS. C. M. BUCKTON.*

FOOD AND HOME COOKERY. With 11 Woodcuts. Crown 8vo. 2s. 6d.

HEALTH IN THE HOUSE. With 41 Woodcuts and Diagrams. Crown 8vo. 2s.

OUR DWELLINGS. With 39 Illustrations. Crown 8vo. 3s. 6d.

Bull.—*WORKS BY THOMAS BULL, M.D.*

HINTS TO MOTHERS ON THE MANAGEMENT OF THEIR HEALTH during the Period of Pregnancy and in the Lying-in Room. Fcp. 8vo. 1s. 6d.

THE MATERNAL MANAGEMENT OF CHILDREN IN HEALTH AND DISEASE. Fcp. 8vo. 1s. 6d.

Bullinger.—*A CRITICAL LEXICON AND CONCORDANCE TO THE ENGLISH AND GREEK NEW TESTAMENT.* Together with an Index of Greek Words and several Appendices. By the Rev. E. W. BULLINGER, D.D. Royal 8vo. 15s.

Burrows.—*THE FAMILY OF BROCAS OF BEAUREPAIRE AND ROCHE COURT,* Hereditary Masters of the Royal Buckhounds. With some account of the English Rule in Aquitaine. By MONTAGU BURROWS, M.A. F.S.A. With 26 Illustrations of Monuments, Brasses, Seals, &c. Royal 8vo. 42s.

Cabinet Lawyer, The; a Popular Digest of the Laws of England, Civil, Criminal, and Constitutional. Fcp. 8vo. 9s.

Canning.—*SOME OFFICIAL CORRESPONDENCE OF GEORGE CANNING.* Edited, with Notes, by EDWARD J. STAPLETON. 2 vols. 8vo. 28s.

Carlyle. — *THOMAS AND JANE WELSH CARLYLE.*

THOMAS CARLYLE, a History of the first Forty Years of his Life, 1795-1835. By J. A. FROUDE, M.A. With 2 Portraits and 4 Illustrations, 2 vols. 8vo. 32s.

THOMAS CARLYLE, a History of his Life in London : from 1834 to his death in 1881. By J. A. FROUDE, M.A. 2 vols. 8vo. 32s.

LETTERS AND MEMORIALS OF JANE WELSH CARLYLE. Prepared for publication by THOMAS CARLYLE, and edited by J. A. FROUDE, M.A. 3 vols. 8vo. 36s.

Cates. — *A DICTIONARY OF GENERAL BIOGRAPHY.* Fourth Edition, with Supplement brought down to the end of 1884. By W. L. R. CATES. 8vo. 28s. cloth ; 35s. half-bound russia.

Clerk.—*THE GAS ENGINE.* By DUGALD CLERK. With 101 Illustrations and Diagrams. Crown 8vo. 7s. 6d.

Coats.—*A MANUAL OF PATHOLOGY.* By JOSEPH COATS, M.D. Pathologist to the Western Infirmary and the Sick Children's Hospital, Glasgow. With 339 Illustrations engraved on Wood. 8vo. 31s. 6d.

Colenso.—*THE PENTATEUCH AND BOOK OF JOSHUA CRITICALLY EXAMINED.* By J. W. COLENSO, D.D. late Bishop of Natal. Crown 8vo. 6s.

Comyn.—*ATHERSTONE PRIORY:* a Tale. By L. N. COMYN. Crown 8vo. 2s. 6d.

Conder. — *A HANDBOOK TO THE BIBLE*, or Guide to the Study of the Holy Scriptures derived from Ancient Monuments and Modern Exploration. By F. R. CONDER, and Lieut. C. R. CONDER, R.E. Post 8vo. 7s. 6d.

Conington. — *WORKS BY JOHN CONINGTON, M.A.*

THE ÆNEID OF VIRGIL. Translated into English Verse. Crown 8vo. 9s.

THE POEMS OF VIRGIL. Translated into English Prose. Crown 8vo. 9s.

Conybeare & Howson. — *THE LIFE AND EPISTLES OF ST. PAUL.* By the Rev. W. J. CONYBEARE, M.A. and the Very Rev. J. S. HOWSON, D.D.

Library Edition, with Maps, Plates, and Woodcuts. 2 vols. square crown 8vo. 21s.

Student's Edition, revised and condensed, with 46 Illustrations and Maps. 1 vol. crown 8vo. 7s. 6d.

Cooke. — *TABLETS OF ANATOMY.* By THOMAS COOKE, F.R.C.S. Eng. B.A. B.Sc. M.D. Paris. Fourth Edition, being a selection of the Tablets believed to be most useful to Students generally. Post 4to. 7s. 6d.

Cox. — *THE FIRST CENTURY OF CHRISTIANITY.* By HOMERSHAM COX, M.A. 8vo. 12s.

Cox.—*A GENERAL HISTORY OF GREECE:* from the Earliest Period to the Death of Alexander the Great; with a Sketch of the History to the Present Time. By the Rev. Sir G. W. COX, Bart., M.A. With 11 Maps and Plans. Crown 8vo. 7s. 6d.

*** For other Works by Sir G. COX, see 'Epochs of History,' p. 24.

Creighton. — *HISTORY OF THE PAPACY DURING THE REFORMATION.* By the Rev. M. CREIGHTON, M.A. 8vo. Vols. I. and II. 1378-1464, 32s.; Vols. III. and IV. 1464-1518, 24s.

Crookes. — *SELECT METHODS IN CHEMICAL ANALYSIS* (chiefly Inorganic). By WILLIAM CROOKES, F.R.S. V.P.C.S. With 37 Illustrations. 8vo. 24s.

Crump.—*A SHORT ENQUIRY INTO THE FORMATION OF POLITICAL OPINION*, from the Reign of the Great Families to the Advent of Democracy. By ARTHUR CRUMP. 8vo. 7s. 6d.

Culley.—*HANDBOOK OF PRACTICAL TELEGRAPHY.* By R. S. CULLEY, M. Inst. C.E. Plates and Woodcuts. 8vo. 16s.

Dante.—*THE DIVINE COMEDY OF DANTE ALIGHIERI.* Translated verse for verse from the Original into Terza Rima. By JAMES INNES MINCHIN. Crown 8vo. 15s.

Davidson.—*AN INTRODUCTION TO THE STUDY OF THE NEW TESTAMENT*, Critical, Exegetical, and Theological. By the Rev. S. DAVIDSON, D.D. LL.D. Revised Edition. 2 vols. 8vo. 30s.

Davidson.—*WORKS BY WILLIAM L. DAVIDSON, M.A.*

THE LOGIC OF DEFINITION EXPLAINED AND APPLIED. Crown 8vo. 6s.

LEADING AND IMPORTANT ENGLISH WORDS EXPLAINED AND EXEMPLIFIED. Fcp. 8vo. 3s. 6d.

Decaisne & Le Maout. — *A GENERAL SYSTEM OF BOTANY.* Translated from the French of E. LE MAOUT, M.D., and J. DECAISNE, by Mrs. HOOKER; with Additions by Sir J. D. HOOKER, C.B. F.R.S. Imp. 8vo. with 5,500 Woodcuts, 31s. 6d.

De Salis. — *WORKS BY MRS. DE SALIS.*

SAVOURIES À LA MODE. Fcp. 8vo. 1s. boards.

ENTRÉES À LA MODE. Fcp. 8vo. 1s. 6d. boards.

De Tocqueville.—*DEMOCRACY IN AMERICA.* By ALEXIS DE TOCQUEVILLE. Translated by HENRY REEVE, C.B. 2 vols. crown 8vo. 16s.

Dickinson. — *ON RENAL AND URINARY AFFECTIONS.* By W. HOWSHIP DICKINSON, M.D. Cantab. F.R.C.P. &c. With 12 Plates and 122 Woodcuts. 3 vols. 8vo. £3. 4s. 6d.

Dixon.—*RURAL BIRD LIFE;* Essays on Ornithology, with Instructions for Preserving Objects relating to that Science. By CHARLES DIXON. With 45 Woodcuts. Crown 8vo. 5s.

Dublin University Press Series

(The) : a Series of Works, chiefly Educational, undertaken by the Provost and Senior Fellows of Trinity College, Dublin :

Abbott's (T. K.) Codex Rescriptus Dublinensis of St. Matthew. 4to. 21*s.*

——————— Evangeliorum Versio Antehieronymiana ex Codice Usseriano (Dublinensi). 2 vols. crown 8vo. 21*s.*

Burnside (W. S.) and Panton's (A. W.) Theory of Equations. 8vo. 12*s.* 6*d.*

Casey's (John) Sequel to Euclid's Elements. Crown 8vo. 3*s.* 6*d.*

——————— Analytical Geometry of the Conic Sections. Crown 8vo. 7*s.* 6*d.*

Davies's (J. F.) Eumenides of Æschylus. With Metrical English Translation. 8vo. 7*s.*

Dublin Translations into Greek and Latin Verse. Edited by R. Y. Tyrrell. 8vo. 12*s.* 6*d.*

Graves's (R. P.) Life of Sir William Hamilton. (3 vols.) Vols. I. and II. 8vo. each 15*s.*

Griffin (R. W.) on Parabola, Ellipse, and Hyperbola, treated Geometrically. Crown 8vo. 6*s.*

Haughton's (Dr. S.) Lectures on Physical Geography. 8vo. 15*s.*

Hobart's (W. K.) Medical Language of St. Luke. 8vo. 16*s.*

Leslie's (T. E. Cliffe) Essays in Political and Moral Philosophy. 8vo. 10*s.* 6*d.*

Macalister's (A.) Zoology and Morphology of Vertebrata. 8vo. 10*s.* 6*d.*

MacCullagh's (James) Mathematical and other Tracts. 8vo. 15*s.*

Maguire's (T.) Parmenides of Plato, Greek Text with English Introduction, Analysis, and Notes. 8vo. 7*s.* 6*d.*

Monck's (W. H. S.) Introduction to Logic. Crown 8vo. 5*s.*

Purser's (J. M.) Manual of Histology. Fcp. 8vo. 5*s.*

Roberts's (R. A.) Examples in the Analytic Geometry of Plane Curves. Fcp. 8vo. 5*s.*

Southey's (R.) Correspondence with Caroline Bowles. Edited by E. Dowden. 8vo. 14*s.*

Thornhill's (W. J.) The Æneid of Virgil, freely translated into English Blank Verse. Crown 8vo. 7*s.* 6*d.*

Tyrrell's (R. Y.) Cicero's Correspondence. Vols. I. and II. 8vo. each 12*s.*

——————— The Acharnians of Aristophanes, translated into English Verse. Crown 8vo. 2*s.* 6*d.*

Webb's (T. E.) Goethe's Faust, Translation and Notes. 8vo. 12*s.* 6*d.*

——————— The Veil of Isis : a Series of Essays on Idealism. 8vo. 10*s.* 6*d.*

Wilkins's (G.) The Growth of the Homeric Poems. 8vo. 6*s.*

Doyle.—*THE OFFICIAL BARONAGE OF ENGLAND.* By JAMES E. DOYLE. Showing the Succession, Dignities, and Offices of every Peer from 1066 to 1885. Vols. I. to III. With 1,600 Portraits, Shields of Arms, Autographs, &c. 3 vols. 4to. £5. 5*s.*

Doyle.—*WORKS BY J. A. DOYLE,* Fellow of All Souls College, Oxford.

THE ENGLISH IN AMERICA : VIRGINIA, MARYLAND, AND THE CAROLINAS. 8vo. 18*s.*

THE ENGLISH IN AMERICA : THE PURITAN COLONIES. 2 vols. 8vo. 36*s.*

Edersheim.—*WORKS BY THE REV. ALFRED EDERSHEIM, D.D.*

THE LIFE AND TIMES OF JESUS THE MESSIAH. 2 vols. 8vo. 24*s.*

PROPHECY AND HISTORY IN RELATION TO THE MESSIAH : the Warburton Lectures, delivered at Lincoln's Inn Chapel, 1880-1884. 8vo. 12*s.*

Ellicott. — *WORKS BY C. J. ELLICOTT, D.D.* Bishop of Gloucester and Bristol.

A CRITICAL AND GRAMMATICAL COMMENTARY ON ST. PAUL'S EPISTLES. 8vo.

I. CORINTHIANS. 16*s.*
GALATIANS. 8*s.* 6*d.*
EPHESIANS. 8*s.* 6*d.*
PASTORAL EPISTLES. 10*s.* 6*d.*
PHILIPPIANS, COLOSSIANS, and PHILEMON. 10*s.* 6*d.*
THESSALONIANS. 7*s.* 6*d.*

HISTORICAL LECTURES ON THE LIFE OF OUR LORD JESUS CHRIST. 8vo. 12*s.*

English Worthies.

Edited by ANDREW LANG, M.A. Fcp. 8vo. 2*s.* 6*d.* each.

DARWIN. By GRANT ALLEN.

MARLBOROUGH. By G. SAINTSBURY.

SHAFTESBURY (The First Earl). By H. D. TRAILL.

ADMIRAL BLAKE. By DAVID HANNAY.

RALEIGH. By EDMUND GOSSE.

STEELE. By AUSTIN DOBSON.

BEN JONSON. By J. A. SYMONDS.

CANNING. By FRANK H. HILL.

CLAVERHOUSE. By MOWBRAY MORRIS.

Epochs of Ancient History.
10 vols. fcp. 8vo. 2s. 6d. each. *See* p. 24.

Epochs of Church History. Fcp.
8vo. 2s. 6d. each. *See* p. 24.

Epochs of English History. . *See*
p. 24.

Epochs of Modern History.
18 vols. fcp. 8vo. 2s. 6d. each. *See* p. 24.

Erichsen.—*WORKS BY JOHN ERIC ERICHSEN, F.R.S.*

THE SCIENCE AND ART OF SURGERY: Being a Treatise on Surgical Injuries, Diseases, and Operations. With 984 Illustrations. 2 vols. 8vo. 42s.

ON CONCUSSION OF THE SPINE, NERVOUS SHOCKS, and other Obscure Injuries of the Nervous System. Cr. 8vo. 10s. 6d.

Ewald. — *WORKS BY PROFESSOR HEINRICH EWALD,* of Göttingen.

THE ANTIQUITIES OF ISRAEL. Translated from the German by H. S. SOLLY, M.A. 8vo. 12s. 6d.

THE HISTORY OF ISRAEL. Translated from the German. 8 vols. 8vo. Vols. I. and II. 24s. Vols. III. and IV. 21s. Vol. V. 18s. Vol. VI. 16s. Vol. VII. 21s. Vol. VIII. with Index to the Complete Work. 18s.

Fairbairn.—*WORKS BY SIR W. FAIRBAIRN, BART. C.E.*

A TREATISE ON MILLS AND MILLWORK, with 18 Plates and 333 Woodcuts. 1 vol. 8vo. 25s.

USEFUL INFORMATION FOR ENGINEERS. With many Plates and Woodcuts. 3 vols. crown 8vo. 31s. 6d.

Farrar. — *LANGUAGE AND LANGUAGES.* A Revised Edition of *Chapters on Language and Families of Speech.* By F. W. FARRAR, D.D. Crown 8vo. 6s.

Firbank.—*THE LIFE AND WORK OF JOSEPH FIRBANK, J.P. D.L.* Railway Contractor. By FREDERICK McDERMOTT, Barrister-at-Law. 8vo. 5s.

Fitzwygram. — *HORSES AND STABLES.* By Major-General Sir F. FITZWYGRAM, Bart. With 19 pages of Illustrations. 8vo. 5s.

Ford.—*THE THEORY AND PRACTICE OF ARCHERY.* By the late HORACE FORD. New Edition, thoroughly Revised and Re-written by W. BUTT, M.A. With a Preface by C. J. LONGMAN, Senior Vice-President Royal Toxophilite Society. 8vo. 14s.

Fox.—*THE EARLY HISTORY OF CHARLES JAMES FOX.* By the Right Hon. Sir G. O. TREVELYAN, Bart.
Library Edition, 8vo. 18s.
Cabinet Edition, cr. 8vo. 6s.

Francis.—*A BOOK ON ANGLING;* or, Treatise on the Art of Fishing in every branch; including full Illustrated Lists of Salmon Flies. By FRANCIS FRANCIS. Post 8vo. Portrait and Plates, 15s.

Freeman.—*THE HISTORICAL GEOGRAPHY OF EUROPE.* By E. A. FREEMAN, D.C.L. With 65 Maps. 2 vols. 8vo. 31s. 6d.

Froude.—*WORKS BY JAMES A. FROUDE, M.A.*

THE HISTORY OF ENGLAND, from the Fall of Wolsey to the Defeat of the Spanish Armada.
Cabinet Edition, 12 vols. cr. 8vo. £3. 12s.
Popular Edition, 12 vols. cr. 8vo. £2. 2s.

SHORT STUDIES ON GREAT SUBJECTS. 4 vols. crown 8vo. 24s.

CÆSAR : a Sketch. Crown 8vo. 6s.

THE ENGLISH IN IRELAND IN THE EIGHTEENTH CENTURY. 3 vols. crown 8vo. 18s.

OCEANA ; OR, ENGLAND AND HER COLONIES. With 9 Illustrations. Crown 8vo. 2s. boards, 2s. 6d. cloth.

THOMAS CARLYLE, a History of the first Forty Years of his Life, 1795 to 1835. 2 vols. 8vo. 32s.

THOMAS CARLYLE, a History of His Life in London from 1834 to his death in 1881. With Portrait engraved on steel. 2 vols. 8vo. 32s.

Ganot. — *WORKS BY PROFESSOR GANOT.* Translated by E. ATKINSON, Ph.D. F.C.S.

ELEMENTARY TREATISE ON PHYSICS. With 5 Coloured Plates and 923 Woodcuts. Crown 8vo. 15s.

NATURAL PHILOSOPHY FOR GENERAL READERS AND YOUNG PERSONS. With 2 Plates, 518 Woodcuts, and an Appendix of Questions. Cr. 8vo. 7s. 6d.

Gardiner. — *WORKS BY SAMUEL RAWSON GARDINER, LL.D.*

HISTORY OF ENGLAND, from the Accession of James I. to the Outbreak of the Civil War, 1603-1642. Cabinet Edition, thoroughly revised. 10 vols. crown 8vo. price 6s. each.

A HISTORY OF THE GREAT CIVIL WAR, 1642-1649. (3 vols.) Vol. I. 1642-1644. With 24 Maps. 8vo. 21s.

OUTLINE OF ENGLISH HISTORY, B.C. 55-A.D. 1880. With 96 Woodcuts, fcp. 8vo. 2s. 6d.

*** For other Works, see 'Epochs of Modern History,' p. 24.

Garrod. — *WORKS BY SIR ALFRED BARING GARROD, M.D. F.R.S.*

A TREATISE ON GOUT AND RHEU-MATIC GOUT (RHEUMATOID ARTHRITIS). With 6 Plates, comprising 21 Figures (14 Coloured), and 27 Illustrations engraved on Wood. 8vo. 21s.

THE ESSENTIALS OF MATERIA MEDICA AND THERAPEUTICS. New Edition, revised and adapted to the New Edition of the British Pharmacopœia, by NESTOR TIRARD, M.D. Crown 8vo. 12s. 6d.

Gilkes. — *BOYS AND MASTERS:* a Story of School Life. By A. H. GILKES, M.A. Head Master of Dulwich College. Crown 8vo. 3s. 6d.

Goethe. — *FAUST.* A New Translation, chiefly in Blank Verse; with Introduction and Notes. By JAMES ADEY BIRDS, B.A. F.G.S. Crown 8vo. 12s. 6d.

FAUST. The German Text, with an English Introduction and Notes for Students. By ALBERT M. SELSS, M.A. Ph.D. Crown 8vo. 5s.

Goodeve. — *WORKS BY T. M. GOOD-EVE, M.A.*

PRINCIPLES OF MECHANICS. With 253 Woodcuts. Crown 8vo. 6s.

THE ELEMENTS OF MECHANISM. With 342 Woodcuts. Crown 8vo. 6s.

A MANUAL OF MECHANICS: an Elementary Text-Book for Students of Applied Mechanics. With 138 Illustrations and Diagrams, and 141 Examples. Fcp. 8vo. 2s. 6d.

Grant. — *THE ETHICS OF ARISTOTLE.* The Greek Text illustrated by Essays and Notes. By Sir ALEXANDER GRANT, Bart. LL.D. D.C.L. &c. 2 vols. 8vo. 32s.

Gray. — *ANATOMY, DESCRIPTIVE AND SURGICAL.* By HENRY GRAY, F.R.S. late Lecturer on Anatomy at St. George's Hospital. With 569 Woodcut Illustrations, a large number of which are coloured. Re-edited by T. PICKERING PICK, Surgeon to St. George's Hospital. Royal 8vo. 36s.

Green. — *THE WORKS OF THOMAS HILL GREEN*, late Fellow of Balliol College, and Whyte's Professor of Moral Philosophy in the University of Oxford. Edited by R. L. NETTLESHIP, Fellow of Balliol College, Oxford (3 vols.) Vols. I. and II.—Philosophical Works. 8vo. 16s. each.

Grove. — *THE CORRELATION OF PHYSICAL FORCES.* By the Hon. Sir W. R. GROVE, F.R.S. &c. 8vo. 15s.

Gwilt. — *AN ENCYCLOPÆDIA OF ARCHITECTURE.* By JOSEPH GWILT, F.S.A. Illustrated with more than 1,100 Engravings on Wood. Revised, with Alterations and Considerable Additions, by WYATT PAPWORTH. 8vo. 52s. 6d.

Haggard. — *WORKS BY H. RIDER HAGGARD.*

SHE: A HISTORY OF ADVENTURE. Crown 8vo. 6s.

ALLAN QUATERMAIN. With 31 Illustrations by C. H. M. KERR. Crown 8vo. 6s.

Halliwell-Phillipps. — *OUTLINES OF THE LIFE OF SHAKESPEARE.* By J. O. HALLIWELL-PHILLIPPS, F.R.S. 2 vols. Royal 8vo. 10s. 6d.

Harte. — *NOVELS BY BRET HARTE.*

IN THE CARQUINEZ WOODS. Fcp. 8vo. 1s. boards; 1s. 6d. cloth.

ON THE FRONTIER. Three Stories. 16mo. 1s.

BY SHORE AND SEDGE. Three Stories. 16mo. 1s.

Hartwig.—*WORKS BY DR. G. HARTWIG.*

THE SEA AND ITS LIVING WONDERS. With 12 Plates and 303 Woodcuts. 8vo. 10*s.* 6*d.*

THE TROPICAL WORLD. With 8 Plates, and 172 Woodcuts. 8vo. 10*s.* 6*d.*

THE POLAR WORLD. With 3 Maps, 8 Plates, and 85 Woodcuts. 8vo. 10*s.* 6*d.*

THE SUBTERRANEAN WORLD. With 3 Maps and 80 Woodcuts. 8vo. 10*s.* 6*d.*

THE AERIAL WORLD. With Map, 8 Plates, and 60 Woodcuts. 8vo. 10*s.* 6*d.*

The following books are extracted from the above works by Dr. HARTWIG :—

DWELLERS IN THE ARCTIC REGIONS. With 29 Illustrations. Crown 8vo. 2*s.* 6*d.* cloth extra, gilt edges.

WINGED LIFE IN THE TROPICS. With 55 Illustrations. Crown 8vo. 2*s.* 6*d.* cloth extra, gilt edges.

VOLCANOES AND EARTHQUAKES. With 30 Illustrations. Crown 8vo. 2*s.* 6*d.* cloth extra, gilt edges.

WILD ANIMALS OF THE TROPICS. With 66 Illustrations. Crown 8vo. 3*s.* 6*d.* cloth extra, gilt edges.

SEA MONSTERS AND SEA BIRDS. With 75 Illustrations. Crown 8vo. 2*s.* 6*d.* cloth extra, gilt edges.

DENIZENS OF THE DEEP. With 117 Illustrations. Crown 8vo. 2*s.* 6*d.* cloth extra, gilt edges.

Hassall.—*THE INHALATION TREATMENT OF DISEASES OF THE ORGANS OF RESPIRATION,* including Consumption. By ARTHUR HILL HASSALL, M.D. With 19 Illustrations of Apparatus. Cr. 8vo. 12*s.* 6*d.*

Havelock. — *MEMOIRS OF SIR HENRY HAVELOCK, K.C.B.* By JOHN CLARK MARSHMAN. Crown 8vo. 3*s.* 6*d.*

Hearn.—*THE GOVERNMENT OF ENGLAND;* its Structure and its Development. By WILLIAM EDWARD HEARN, Q.C. 8vo. 16*s.*

Helmholtz. — *WORKS BY PROFESSOR HELMHOLTZ.*

ON THE SENSATIONS OF TONE AS A PHYSIOLOGICAL BASIS FOR THE THEORY OF MUSIC. Royal 8vo. 28*s.*

POPULAR LECTURES ON SCIENTIFIC SUBJECTS. With 68 Woodcuts. 2 vols. Crown 8vo. 15*s.* or separately, 7*s.* 6*d.* each.

Herschel.—*OUTLINES OF ASTRONOMY.* By Sir J. F. W. HERSCHEL, Bart. M.A. With Plates and Diagrams. Square crown 8vo. 12*s.*

Hester's Venture : a Novel. By the Author of 'The Atelier du Lys.' Crown 8vo. 2*s.* 6*d.*

Hewitt. — *THE DIAGNOSIS AND TREATMENT OF DISEASES OF WOMEN, INCLUDING THE DIAGNOSIS OF PREGNANCY.* By GRAILY HEWITT, M.D. With 211 Engravings. 8vo. 24*s.*

Historic Towns. Edited by E. A. FREEMAN, D.C.L. and Rev. WILLIAM HUNT, M.A. With Maps and Plans. Crown 8vo. 3*s.* 6*d.* each.

LONDON. By W. E. LOFTIE.

EXETER. By E. A. FREEMAN.

BRISTOL. By W. HUNT.

OXFORD. By C. W. BOASE.

[*₊* Other Volumes are in preparation.

Hobart.—*SKETCHES FROM MY LIFE.* By Admiral HOBART PASHA. With Portrait. Crown 8vo. 7*s.* 6*d.*

Holmes.—*A SYSTEM OF SURGERY,* Theoretical and Practical, in Treatises by various Authors. Edited by TIMOTHY HOLMES, M.A. and J. W. HULKE, F.R.S. 3 vols. royal 8vo. £4. 4*s.*

Homer.—*THE ILIAD OF HOMER,* Homometrically translated by C. B. CAYLEY. 8vo. 12*s.* 6*d.*

THE ILIAD OF HOMER. The Greek Text, with a Verse Translation, by W. C. GREEN, M.A. Vol. I. Books I.–XII. Crown 8vo. 6*s.*

Hopkins.—*CHRIST THE CONSOLER;* a Book of Comfort for the Sick. By ELLICE HOPKINS. Fcp. 8vo. 2*s.* 6*d.*

Howitt.—*VISITS TO REMARKABLE PLACES,* Old Halls, Battle-Fields, Scenes illustrative of Striking Passages in English History and Poetry. By WILLIAM HOWITT. With 80 Illustrations engraved on Wood. Crown 8vo. 7*s.* 6*d.*

Hudson & Gosse.—*THE ROTIFERA OR 'WHEEL-ANIMALCULES.'* By C. T. HUDSON, LL.D. and P. H. GOSSE, F.R.S. With 30 Coloured Plates. In 6 Parts. 4to. 10*s.* 6*d.* each. Complete in 2 vols. 4to. £3. 10*s.*

Hullah.—*WORKS BY JOHN HUL-LAH, LL.D.*

COURSE OF LECTURES ON THE HISTORY OF MODERN MUSIC. 8vo. 8s. 6d.

COURSE OF LECTURES ON THE TRANSITION PERIOD OF MUSICAL HISTORY. 8vo. 10s. 6d.

Hume.—*THE PHILOSOPHICAL WORKS OF DAVID HUME.* Edited by T. H. GREEN, M.A. and the Rev. T. H. GROSE, M.A. 4 vols. 8vo. 56s. Or separately, Essays, 2 vols. 28s. Treatise of Human Nature. 2 vols. 28s.

Huth.—*THE MARRIAGE OF NEAR KIN,* considered with respect to the Law of Nations, the Result of Experience, and the Teachings of Biology. By ALFRED H. HUTH. Royal 8vo. 21s.

In the Olden Time: a Tale of the Peasant War in Germany. By the Author of 'Mademoiselle Mori.' Crown 8vo. 2s. 6d.

Ingelow.—*WORKS BY JEAN INGELOW.*

POETICAL WORKS. Vols. 1 and 2. Fcp. 8vo. 12s.

LYRICAL AND OTHER POEMS. Selected from the Writings of JEAN INGELOW. Fcp. 8vo. 2s. 6d. cloth plain; 3s. cloth gilt.

Jackson.—*AID TO ENGINEERING SOLUTION.* By LOWIS D'A. JACKSON, C.E. With 111 Diagrams and 5 Woodcut Illustrations. 8vo. 21s.

Jameson.—*WORKS BY MRS. JAMESON.*

LEGENDS OF THE SAINTS AND MARTYRS. With 19 Etchings and 187 Woodcuts. 2 vols. 31s. 6d.

LEGENDS OF THE MADONNA, the Virgin Mary as represented in Sacred and Legendary Art. With 27 Etchings and 165 Woodcuts. 1 vol. 21s.

LEGENDS OF THE MONASTIC ORDERS. With 11 Etchings and 88 Woodcuts. 1 vol. 21s.

HISTORY OF THE SAVIOUR, His Types and Precursors. Completed by Lady EASTLAKE. With 13 Etchings and 281 Woodcuts. 2 vols. 42s.

Jeans.—*WORKS BY J. S. JEANS.*

ENGLAND'S SUPREMACY: its Sources, Economics, and Dangers. 8vo. 8s. 6d.

RAILWAY PROBLEMS: An Inquiry into the Economic Conditions of Railway Working in Different Countries. 8vo. 12s. 6d.

Johnson.—*THE PATENTEE'S MANUAL;* a Treatise on the Law and Practice of Letters Patent. By J. JOHNSON and J. H. JOHNSON. 8vo. 10s. 6d.

Johnston.—*A GENERAL DICTIONARY OF GEOGRAPHY,* Descriptive, Physical, Statistical, and Historical; a complete Gazetteer of the World. By KEITH JOHNSTON. Medium 8vo. 42s.

Johnstone.—*A SHORT INTRODUCTION TO THE STUDY OF LOGIC.* By LAURENCE JOHNSTONE. Crown 8vo. 2s. 6d.

Jordan. — *WORKS BY WILLIAM LEIGHTON JORDAN, F.R.G.S.*

THE OCEAN: a Treatise on Ocean Currents and Tides and their Causes. 8vo. 21s.

THE NEW PRINCIPLES OF NATURAL PHILOSOPHY. With 13 plates. 8vo. 21s.

THE WINDS: an Essay in Illustration of the New Principles of Natural Philosophy. Crown 8vo. 2s.

THE STANDARD OF VALUE. Crown 8vo. 5s.

Jukes.—*WORKS BY ANDREW JUKES.*

THE NEW MAN AND THE ETERNAL LIFE. Crown 8vo. 6s.

THE TYPES OF GENESIS. Crown 8vo. 7s. 6d.

THE SECOND DEATH AND THE RESTITUTION OF ALL THINGS. Crown 8vo. 3s. 6d.

THE MYSTERY OF THE KINGDOM. Crown 8vo. 2s. 6d.

Justinian. — *THE INSTITUTES OF JUSTINIAN;* Latin Text, chiefly that of Huschke, with English Introduction, Translation, Notes, and Summary. By THOMAS C. SANDARS, M.A. 8vo. 18s.

Kalisch. — *WORKS BY M. M. KALISCH, M.A.*

BIBLE STUDIES. Part I. The Prophecies of Balaam. 8vo. 10s. 6d. Part II. The Book of Jonah. 8vo. 10s. 6d.

COMMENTARY ON THE OLD TESTAMENT; with a New Translation. Vol. I. Genesis, 8vo. 18s. or adapted for the General Reader, 12s. Vol. II. Exodus, 15s. or adapted for the General Reader, 12s. Vol. III. Leviticus, Part I. 15s. or adapted for the General Reader, 8s. Vol. IV. Leviticus, Part II. 15s. or adapted for the General Reader, 8s.

HEBREW GRAMMAR. With Exercises. Part I. 8vo. 12s. 6d. Key, 5s. Part II. 12s. 6d.

Kant.—*WORKS BY EMMANUEL KANT.*

CRITIQUE OF PRACTICAL REASON. Translated by Thomas Kingsmill Abbott, B.D. 8vo. 12s. 6d.

INTRODUCTION TO LOGIC, AND HIS ESSAY ON THE MISTAKEN SUBTILTY OF THE FOUR FIGURES. Translated by Thomas Kingsmill Abbott, B.D. With a few Notes by S. T. Coleridge. 8vo. 6s.

Kendall.—*WORKS BY MAY KENDALL.*

FROM A GARRET. Crown 8vo. 6s.

DREAMS TO SELL; Poems. Crown 8vo. 6s.

Killick.—*HANDBOOK TO MILL'S SYSTEM OF LOGIC.* By the Rev. A. H. KILLICK, M.A. Crown 8vo. 3s. 6d.

Kirkup.—*AN INQUIRY INTO SOCIALISM.* By THOMAS KIRKUP, Author of the Article on ' Socialism ' in the ' Encyclopædia Britannica.' Crown 8vo. 5s.

Knowledge Library. (*See* PROCTOR'S Works, p. 16.)

Kolbe.—*A SHORT TEXT-BOOK OF INORGANIC CHEMISTRY.* By Dr. HERMANN KOLBE. Translated from the German by T. S. HUMPIDGE, Ph.D. With a Coloured Table of Spectra and 66 Illustrations. Crown 8vo. 7s. 6d.

Ladd. — *ELEMENTS OF PHYSIOLOGICAL PSYCHOLOGY:* a Treatise of the Activities and Nature of the Mind from the Physical and Experimental Point of View. By GEORGE T. LADD. With 113 Illustrations and Diagrams. 8vo. 21s.

Lang.—*WORKS BY ANDREW LANG.*

MYTH, RITUAL, AND RELIGION. 2 vols. crown 8vo. 21s.

CUSTOM AND MYTH; Studies of Early Usage and Belief. With 15 Illustrations. Crown 8vo. 7s. 6d.

LETTERS TO DEAD AUTHORS. Fcp. 8vo. 6s. 6d.

BOOKS AND BOOKMEN. With 2 Coloured Plates and 17 Illustrations. Cr. 8vo. 6s. 6d.

JOHNNY NUT AND THE GOLDEN GOOSE. Done into English by ANDREW LANG, from the French of CHARLES DEULIN. Illustrated by Am. Lynen. Royal 8vo. 10s. 6d. gilt edges.

Larden.—*ELECTRICITY FOR PUBLIC SCHOOLS AND COLLEGES.* With numerous Questions and Examples with Answers, and 214 Illustrations and Diagrams. By W. LARDEN, M.A. Crown 8vo. 6s.

Laughton.—*STUDIES IN NAVAL HISTORY;* Biographies. By J. K. LAUGHTON, M.A. Professor of Modern History at King's College, London. 8vo. 10s. 6d.

Lecky.—*WORKS BY W. E. H. LECKY.*

HISTORY OF ENGLAND IN THE EIGHTEENTH CENTURY. 8vo. Vols. I. & II. 1700–1760. 36s. Vols. III. & IV. 1760–1784. 36s. Vols. V. & VI. 1784–1793. 36s.

THE HISTORY OF EUROPEAN MORALS FROM AUGUSTUS TO CHARLEMAGNE. 2 vols. crown 8vo. 16s.

HISTORY OF THE RISE AND INFLUENCE OF THE SPIRIT OF RATIONALISM IN EUROPE. 2 vols. crown 8vo. 16s.

Lewes.—*THE HISTORY OF PHILOSOPHY,* from Thales to Comte. By GEORGE HENRY LEWES. 2 vols. 8vo. 32s.

Lindt.—*PICTURESQUE NEW GUINEA.* By J. W. LINDT, F.R.G.S. With 50 Full-page Photographic Illustrations reproduced by the Autotype Company. Crown 4to. 42s.

Liveing.—*WORKS BY ROBERT LIVEING, M.A. and M.D. Cantab.*

HANDBOOK ON DISEASES OF THE SKIN. With especial reference to Diagnosis and Treatment. Fcp 8vo. 5s.

NOTES ON THE TREATMENT OF SKIN DISEASES. 18mo. 3s.

Lloyd.—*A TREATISE ON MAGNETISM,* General and Terrestrial. By H. LLOYD, D.D. D.C.L. 8vo. 10s. 6d.

Lloyd.—*THE SCIENCE OF AGRICULTURE.* By F. J. LLOYD. 8vo. 12s.

Longman.—*HISTORY OF THE LIFE AND TIMES OF EDWARD III.* By WILLIAM LONGMAN, F.S.A. With 9 Maps, 8 Plates, and 16 Woodcuts. 2 vols. 8vo. 28s.

Longman.—*WORKS BY FREDERICK W. LONGMAN, Balliol College, Oxon.*

CHESS OPENINGS. Fcp. 8vo. 2s. 6d.

FREDERICK THE GREAT AND THE SEVEN YEARS' WAR. With 2 Coloured Maps. 8vo. 2s. 6d.

A NEW POCKET DICTIONARY OF THE GERMAN AND ENGLISH LANGUAGES. Square 18mo. 2s. 6d.

Longman's Magazine. Published Monthly. Price Sixpence.
Vols. 1-10, 8vo. price 5*s.* each.

Longmore.— *GUNSHOT INJURIES*; Their History, Characteristic Features, Complications, and General Treatment. By Surgeon-General Sir T. LONGMORE, C.B., F.R.C.S. With 58 Illustrations. 8vo. 31*s.* 6*d.*

Loudon.— *WORKS BY J. C. LOUDON, F.L.S.*

ENCYCLOPÆDIA OF GARDENING; the Theory and Practice of Horticulture, Floriculture, Arboriculture, and Landscape Gardening. With 1,000 Woodcuts. 8vo. 21*s.*

ENCYCLOPÆDIA OF AGRICULTURE; the Laying-out, Improvement, and Management of Landed Property; the Cultivation and Economy of the Productions of Agriculture. With 1,100 Woodcuts. 8vo. 21*s.*

ENCYCLOPÆDIA OF PLANTS; the Specific Character, Description, Culture, History, &c. of all Plants found in Great Britain. With 12,000 Woodcuts. 8vo. 42*s.*

Lubbock.— *THE ORIGIN OF CIVILIZATION AND THE PRIMITIVE CONDITION OF MAN.* By Sir J. LUBBOCK, Bart. M.P. F.R.S. With Illustrations. 8vo. 18*s.*

Lyall.— *THE AUTOBIOGRAPHY OF A SLANDER.* By EDNA LYALL, Author of 'Donovan,' 'We Two,' &c. Fcp. 8vo. 1*s.* sewed.

Lyra Germanica; Hymns Translated from the German by Miss C. WINKWORTH. Fcp. 8vo. 5*s.*

Macaulay.— *WORKS AND LIFE OF LORD MACAULAY.*

HISTORY OF ENGLAND FROM THE ACCESSION OF JAMES THE SECOND:
Student's Edition, 2 vols. crown 8vo. 12*s.*
People's Edition, 4 vols. crown 8vo. 16*s.*
Cabinet Edition, 8 vols. post 8vo. 48*s.*
Library Edition, 5 vols. 8vo. £4.

CRITICAL AND HISTORICAL ESSAYS, with LAYS of ANCIENT ROME, in 1 volume:
Authorised Edition, crown 8vo. 2*s.* 6*d.* or 3*s.* 6*d.* gilt edges.
Popular Edition, crown 8vo. 2*s.* 6*d.*

CRITICAL AND HISTORICAL ESSAYS:
Student's Edition, 1 vol. crown 8vo. 6*s.*
People's Edition, 2 vols. crown 8vo. 8*s.*
Cabinet Edition, 4 vols. post 8vo. 24*s.*
Library Edition, 3 vols. 8vo. 36*s.*

Macaulay— *WORKS AND LIFE OF LORD MACAULAY—continued.*

ESSAYS which may be had separately price 6*d.* each sewed, 1*s.* each cloth:
Addison and Walpole.
Frederick the Great.
Croker's Boswell's Johnson.
Hallam's Constitutional History.
Warren Hastings. (3*d.* sewed, 6*d.* cloth.)
The Earl of Chatham (Two Essays).
Ranke and Gladstone.
Milton and Machiavelli.
Lord Bacon.
Lord Clive.
Lord Byron, and The Comic Dramatists of the Restoration.

The Essay on Warren Hastings annotated by S. HALES, 1*s.* 6*d.*
The Essay on Lord Clive annotated by H. COURTHOPE BOWEN, M.A. 2*s.* 6*d.*

SPEECHES:
People's Edition, crown 8vo. 3*s.* 6*d.*

MISCELLANEOUS WRITINGS:
Library Edition, 2 vols. 8vo. 21*s.*
People's Edition, 1 vol. crown 8vo. 4*s.* 6*d.*

LAYS OF ANCIENT ROME, &c.
Illustrated by G. Scharf, fcp. 4to. 10*s.* 6*d.*
————————— Popular Edition, fcp. 4to. 6*d.* sewed, 1*s.* cloth.
Illustrated by J. R. Weguelin, crown 8vo. 3*s.* 6*d.* cloth extra, gilt edges.
Cabinet Edition, post 8vo. 3*s.* 6*d.*
Annotated Edition, fcp. 8vo. 1*s.* sewed, 1*s.* 6*d.* cloth, or 2*s.* 6*d.* cloth extra, gilt edges.

SELECTIONS FROM THE WRITINGS OF LORD MACAULAY. Edited, with Occasional Notes, by the Right Hon. Sir G. O. TREVELYAN, Bart. Crown 8vo. 6*s.*

MISCELLANEOUS WRITINGS AND SPEECHES:
Student's Edition, in ONE VOLUME, crown 8vo. 6*s.*
Cabinet Edition, including Indian Penal Code, Lays of Ancient Rome, and Miscellaneous Poems, 4 vols. post 8vo. 24*s.*

THE COMPLETE WORKS OF LORD MACAULAY. Edited by his Sister, Lady TREVELYAN.
Library Edition, with Portrait, 8 vols. demy 8vo. £5. 5*s.*
Cabinet Edition, 16 vols. post 8vo. £4. 16*s.*

THE LIFE AND LETTERS OF LORD MACAULAY. By the Right Hon. Sir G. O. TREVELYAN, Bart.
Popular Edition, 1 vol. crown 8vo. 6*s.*
Cabinet Edition, 2 vols. post 8vo. 12*s.*
Library Edition, 2 vols. 8vo. 36*s.*

Macdonald.—*WORKS BY GEORGE MACDONALD, LL.D.*

UNSPOKEN SERMONS. First Series. Crown 8vo. 3s. 6d.

UNSPOKEN SERMONS. Second Series. Crown 8vo. 3s. 6d.

THE MIRACLES OF OUR LORD. Crown 8vo. 3s. 6d.

A BOOK OF STRIFE, IN THE FORM OF THE DIARY OF AN OLD SOUL: Poems. 12mo. 6s.

Macfarren.—*LECTURES ON HAR-MONY,* delivered at the Royal Institution. By Sir G. A. MACFARREN. 8vo. 12s.

Macleod.—*WORKS BY HENRY D. MACLEOD, M.A.*

THE ELEMENTS OF ECONOMICS. In 2 vols. Vol. I. crown 8vo. 7s. 6d. Vol. II. PART 1, crown 8vo. 7s. 6d.

THE ELEMENTS OF BANKING. Crown 8vo. 5s.

THE THEORY AND PRACTICE OF BANKING. Vol. I. 8vo. 12s. Vol. II. 14s.

McCulloch. — *THE DICTIONARY OF COMMERCE AND COMMERCIAL NAVI-GATION* of the late J. R. McCULLOCH, of H.M. Stationery Office. Latest Edition, containing the most recent Statistical Information by A. J. WILSON. 1 vol. medium 8vo. with 11 Maps and 30 Charts, price 63s. cloth, or 70s. strongly half-bound in russia.

Mademoiselle Mori: a Tale of Modern Rome. By the Author of 'The Atelier du Lys.' Crown 8vo. 2s. 6d.

Mahaffy.—*A HISTORY OF CLAS-SICAL GREEK LITERATURE.* By the Rev. J. P. MAHAFFY, M.A. Crown 8vo. Vol. I. Poets, 7s. 6d. Vol. II. Prose Writers, 7s. 6d.

Malmesbury. — *MEMOIRS OF AN EX-MINISTER:* an Autobiography. By the Earl of MALMESBURY, G.C.B. Crown 8vo. 7s. 6d.

Manning.—*THE TEMPORAL MIS-SION OF THE HOLY GHOST;* or, Reason and Revelation. By H. E. MANNING, D.D. Cardinal-Archbishop. Crown 8vo. 8s. 6d.

Martineau—*WORKS BY JAMES MARTINEAU, D.D.*

HOURS OF THOUGHT ON SACRED THINGS. Two Volumes of Sermons. 2 vols. crown 8vo. 7s. 6d. each.

ENDEAVOURS AFTER THE CHRISTIAN LIFE. Discourses. Crown 8vo. 7s. 6d.

Maunder's Treasuries.

BIOGRAPHICAL TREASURY. Recon-structed, revised, and brought down to the year 1882, by W. L. R. CATES. Fcp. 8vo. 6s.

TREASURY OF NATURAL HISTORY; or, Popular Dictionary of Zoology. Fcp. 8vo. with 900 Woodcuts, 6s.

TREASURY OF GEOGRAPHY, Physical, Historical, Descriptive, and Political. With 7 Maps and 16 Plates. Fcp. 8vo. 6s.

HISTORICAL TREASURY: Outlines of Universal History, Separate Histories of all Nations. Revised by the Rev. Sir G. W. Cox, Bart. M.A. Fcp. 8vo. 6s.

TREASURY OF KNOWLEDGE AND LIBRARY OF REFERENCE. Comprising an English Dictionary and Grammar, Universal Gazetteer, Classical Dictionary, Chronology, Law Dictionary, &c. Fcp. 8vo. 6s.

SCIENTIFIC AND LITERARY TREA-SURY: a Popular Encyclopædia of Science, Literature, and Art. Fcp. 8vo. 6s.

THE TREASURY OF BIBLE KNOW-LEDGE; being a Dictionary of the Books, Persons, Places, Events, and other matters of which mention is made in Holy Scrip-ture. By the Rev. J. AYRE, M.A. With 5 Maps, 15 Plates, and 300 Woodcuts. Fcp. 8vo. 6s.

THE TREASURY OF BOTANY, or Popular Dictionary of the Vegetable Kingdom. Edited by J. LINDLEY, F.R.S. and T. MOORE, F.L.S. With 274 Wood-cuts and 20 Steel Plates. Two Parts, fcp. 8vo. 12s.

May.—*WORKS BY THE RIGHT HON. SIR THOMAS ERSKINE MAY, K.C.B*

THE CONSTITUTIONAL HISTORY OF ENGLAND SINCE THE ACCESSION OF GEORGE III. 1760-1870. 3 vols. crown 8vo. 18s.

DEMOCRACY IN EUROPE; a History. 2 vols. 8vo. 32s.

Melville.—*NOVELS BY G. J. WHYTE MELVILLE.* Crown 8vo. 1s. each, boards; 1s. 6d. each, cloth.

The Gladiators.	Holmby House.
The Interpreter.	Kate Coventry.
Good for Nothing.	Digby Grand.
The Queen's Maries.	General Bounce.

Mendelssohn.—*THE LETTERS OF FELIX MENDELSSOHN.* Translated by Lady WALLACE. 2 vols. crown 8vo. 10s.

Merivale.— *Works by the Very Rev. Charles Merivale, D.D. Dean of Ely.*

History of the Romans under the Empire. 8 vols. post 8vo. 48*s.*

The Fall of the Roman Republic: a Short History of the Last Century tury of the Commonwealth. 12mo. 7*s.* 6*d.*

General History of Rome from b.c. 753 *to a.d.* 476. Crown 8vo. 7*s.* 6*d.*

The Roman Triumvirates. With Maps. Fcp. 8vo. 2*s.* 6*d.*

Mill.— *Analysis of the Phenomena of the Human Mind.* By James Mill. With Notes, Illustrative and Critical. 2 vols. 8vo. 28*s.*

Mill.— *Works by John Stuart Mill.*

Principles of Political Economy. Library Edition, 2 vols. 8vo. 30*s.* People's Edition, 1 vol. crown 8vo. 5*s.*

A System of Logic, Ratiocinative and Inductive. Crown 8vo. 5*s.*

On Liberty. Crown 8vo. 1*s.* 4*d.*

On Representative Government. Crown 8vo. 2*s.*

Autobiography. 8vo. 7*s.* 6*d.*

Utilitarianism. 8vo. 5*s.*

Examination of Sir William Hamilton's Philosophy. 8vo. 16*s.*

Nature, the Utility of Religion, and Theism. Three Essays. 8vo. 5*s.*

Miller.— *Works by W. Allen Miller, M.D. LL.D.*

The Elements of Chemistry, Theoretical and Practical. Re-edited, with Additions, by H. Macleod, F.C.S. 3 vols. 8vo.
Vol. I. Chemical Physics, 16*s.*
Vol. II. Inorganic Chemistry, 24*s.*
Vol. III. Organic Chemistry, 31*s.* 6*d.*

An Introduction to the Study of Inorganic Chemistry. With 71 Woodcuts. Fcp. 8vo. 3*s.* 6*d.*

Mitchell.— *A Manual of Practical Assaying.* By John Mitchell, F.C.S. Revised, with the Recent Discoveries incorporated. By W. Crookes, F.R.S. 8vo. Woodcuts, 31*s.* 6*d.*

Molesworth. — *Marrying and Giving in Marriage:* a Novel. By Mrs. Molesworth. Fcp. 8vo. 2*s.* 6*d.*

Monsell.— *Works by the Rev. J. S. B. Monsell, LL.D.*

Spiritual Songs for the Sundays and Holydays throughout the Year. Fcp. 8vo. 5*s.* 18mo. 2*s.*

The Beatitudes. Eight Sermons. Crown 8vo. 3*s.* 6*d.*

His Presence not His Memory. Verses. 16mo. 1*s.*

Mulhall.— *History of Prices since the Year* 1850. By Michael G. Mulhall. Crown 8vo. 6*s.*

Müller. — *Works by F. Max Müller, M.A.*

Biographical Essays. Crown 8vo. 7*s.* 6*d.*

Selected Essays on Language, Mythology and Religion. 2 vols. crown 8vo. 16*s.*

Lectures on the Science of Language. 2 vols. crown 8vo. 16*s.*

India, What Can it Teach Us? A Course of Lectures delivered before the University of Cambridge. 8vo. 12*s.* 6*d.*

Hibbert Lectures on the Origin and Growth of Religion, as illustrated by the Religions of India. Crown 8vo. 7*s.* 6*d.*

Introduction to the Science of Religion: Four Lectures delivered at the Royal Institution. Crown 8vo. 7*s.* 6*d.*

The Science of Thought. 8vo. 21*s.*

A Sanskrit Grammar for Beginners. New and Abridged Edition, accented and transliterated throughout, with a chapter on Syntax and an Appendix on Classical Metres. By A. A. MacDonell, M.A. Ph.D. Crown 8vo. 6*s.*

Munk.— *Euthanasia;* or, Medical Treatment in Aid of an Easy Death. By William Munk, M.D. F.S.A. Fellow and late Senior Censor of the Royal College of Physicians, &c. Crown 8vo. 4*s.* 6*d.*

Murchison.—*WORKS BY CHARLES MURCHISON, M.D. LL.D. &c.*

A TREATISE ON THE CONTINUED FEVERS OF GREAT BRITAIN. Revised by W. CAYLEY, M.D. Physician to the Middlesex Hospital. 8vo. with numerous Illustrations, 25*s.*

CLINICAL LECTURES ON DISEASES OF THE LIVER, JAUNDICE, AND ABDOMINAL DROPSY. Revised by T. LAUDER BRUNTON, M.D. and Sir JOSEPH FAYRER, M.D. 8vo. with 43 Illustrations, 24*s.*

Napier.—*THE LIFE OF SIR JOSEPH NAPIER, BART. EX-LORD CHANCELLOR OF IRELAND.* From his Private Correspondence. By ALEX. CHARLES EWALD, F.S.A. With Portrait on Steel, engraved by G. J. Stodart, from a Photograph. 8vo. 15*s.*

Nelson.—*LETTERS AND DESPATCHES OF HORATIO, VISCOUNT NELSON.* Selected and arranged by JOHN KNOX LAUGHTON, M.A. 8vo. 16*s.*

Nesbit.—*LAYS AND LEGENDS.* By E. NESBIT. Crown 8vo. 5*s.*

Newman.—*WORKS BY CARDINAL NEWMAN.*

APOLOGIA PRO VITA SUA. Crown 8vo. 6*s.*

THE IDEA OF A UNIVERSITY DEFINED AND ILLUSTRATED. Crown 8vo. 7*s.*

HISTORICAL SKETCHES. 3 vols. crown 8vo. 6*s.* each.

DISCUSSIONS AND ARGUMENTS ON VARIOUS SUBJECTS. Crown 8vo. 6*s.*

AN ESSAY ON THE DEVELOPMENT OF CHRISTIAN DOCTRINE. Crown 8vo. 6*s.*

CERTAIN DIFFICULTIES FELT BY ANGLICANS IN CATHOLIC TEACHING CONSIDERED. Vol. 1, crown 8vo. 7*s.* 6*d.*; Vol. 2, crown 8vo. 5*s.* 6*d.*

THE VIA MEDIA OF THE ANGLICAN CHURCH, ILLUSTRATED IN LECTURES &c. 2 vols. crown 8vo. 6*s.* each.

ESSAYS, CRITICAL AND HISTORICAL. 2 vols. crown 8vo. 12*s.*

ESSAYS ON BIBLICAL AND ON ECCLESIASTICAL MIRACLES. Crown 8vo. 6*s.*

AN ESSAY IN AID OF A GRAMMAR OF ASSENT. 7*s.* 6*d.*

Noble.—*HOURS WITH A THREE-INCH TELESCOPE.* By Captain W. NOBLE, F.R.A.S. &c. With a Map of the Moon. Crown 8vo. 4*s.* 6*d.*

Northcott.—*LATHES AND TURNING,* Simple, Mechanical, and Ornamental. By W. H. NORTHCOTT. With 338 Illustrations. 8vo. 18*s.*

Oliphant.—*NOVELS BY MRS. OLIPHANT.*

MADAM. Crown 8vo. 1*s.* boards; 1*s.* 6*d.* cloth.

IN TRUST.—Crown 8vo. 1*s.* boards; 1*s.* 6*d.* cloth.

Oliver. — *ASTRONOMY FOR AMATEURS:* a Practical Manual of Telescopic Research adapted to Moderate Instruments. Edited by J. A. WESTWOOD OLIVER, with the assistance of Messrs. MAUNDER, GRUBB, GORE, DENNING, FRANKS, ELGER, BURNHAM, CAPRON, BACKHOUSE, and others. With several Illustrations. Crown 8vo. 7*s.* 6*d.*

Overton.—*LIFE IN THE ENGLISH CHURCH* (1660-1714). By J. H. OVERTON, M.A. Rector of Epworth. 8vo. 14*s.*

Owen. — *THE COMPARATIVE ANATOMY AND PHYSIOLOGY OF THE VERTEBRATE ANIMALS.* By Sir RICHARD OWEN, K.C.B. &c. With 1,472 Woodcuts. 3 vols. 8vo. £3. 13*s.* 6*d.*

Paget. — *WORKS BY SIR JAMES PAGET, BART. F.R.S. D.C.L. &c.*

CLINICAL LECTURES AND ESSAYS. Edited by F. HOWARD MARSH, Assistant-Surgeon to St. Bartholomew's Hospital. 8vo. 15*s.*

LECTURES ON SURGICAL PATHOLOGY. Re-edited by the AUTHOR and W. TURNER, M.B. 8vo. with 131 Woodcuts, 21*s.*

Pasteur.—*LOUIS PASTEUR,* his Life and Labours. By his SON-IN-LAW. Translated from the French by Lady CLAUD HAMILTON. Crown 8vo. 7*s.* 6*d.*

Payen.—*INDUSTRIAL CHEMISTRY;* a Manual for Manufacturers and for Colleges or Technical Schools; a Translation of PAYEN'S 'Précis de Chimie Industrielle.' Edited by B. H. PAUL. With 698 Woodcuts. Medium 8vo. 42*s.*

Payn.—*NOVELS BY JAMES PAYN.*

THE LUCK OF THE DARRELLS. Crown 8vo. 1*s.* boards; 1*s.* 6*d.* cloth.

THICKER THAN WATER. Crown 8vo. 1*s.* boards; 1*s.* 6*d.* cloth.

Pears.—*THE FALL OF CONSTANTINOPLE:* being the Story of the Fourth Crusade. By EDWIN PEARS, LL.B. Barrister-at-Law, late President of the European Bar at Constantinople, and Knight of the Greek Order of the Saviour. 8vo. 16*s.*

Perring.—*HARD KNOTS IN SHAKES-PEARE.* By Sir PHILIP PERRING, Bart. 8vo. 7*s.* 6*d.*

Piesse.—*THE ART OF PERFUMERY,* and the Methods of Obtaining the Odours of Plants; with Instructions for the Manufacture of Perfumes, &c. By G. W. S. PIESSE, Ph.D. F.C.S. With 96 Woodcuts, square crown 8vo. 21*s.*

Pole.—*THE THEORY OF THE MO-DERN SCIENTIFIC GAME OF WHIST.* By W. POLE, F.R.S. Fcp. 8vo. 2*s.* 6*d.*

Prendergast.—*IRELAND,* from the Restoration to the Revolution, 1660-1690. By JOHN P. PRENDERGAST. 8vo. 5*s.*

Proctor.—*WORKS BY R. A. PROC-TOR.*

THE ORBS AROUND US; a Series of Essays on the Moon and Planets, Meteors and Comets. With Chart and Diagrams, crown 8vo. 5*s.*

OTHER WORLDS THAN OURS; The Plurality of Worlds Studied under the Light of Recent Scientific Researches. With 14 Illustrations, crown 8vo. 5*s.*

THE MOON; her Motions, Aspects, Scenery, and Physical Condition. With Plates, Charts, Woodcuts, and Lunar Photographs, crown 8vo. 6*s.*

UNIVERSE OF STARS; Presenting Researches into and New Views respecting the Constitution of the Heavens. With 22 Charts and 22 Diagrams, 8vo. 10*s.* 6*d.*

LARGER STAR ATLAS for the Library, in 12 Circular Maps, with Introduction and 2 Index Pages. Folio, 15*s.* or Maps only, 12*s.* 6*d.*

NEW STAR ATLAS for the Library, the School, and the Observatory, in 12 Circular Maps (with 2 Index Plates). Crown 8vo. 5*s.*

LIGHT SCIENCE FOR LEISURE HOURS; Familiar Essays on Scientific Subjects, Natural Phenomena, &c. 3 vols. crown 8vo. 5*s.* each.

CHANCE AND LUCK; a Discussion of the Laws of Luck, Coincidences, Wagers, Lotteries, and the Fallacies of Gambling &c. Crown 8vo. 5*s.*

STUDIES OF VENUS-TRANSITS; an Investigation of the Circumstances of the Transits of Venus in 1874 and 1882. With 7 Diagrams and 10 Plates. 8vo. 5*s.*

The **'KNOWLEDGE' LIBRARY.** Edited by RICHARD A. PROCTOR.

HOW TO PLAY WHIST: WITH THE LAWS AND ETIQUETTE OF WHIST. By R. A. PROCTOR. Crown 8vo. 5*s.*

HOME WHIST: an Easy Guide to Correct Play. By R. A. PROCTOR. 16mo. 1*s.*

THE POETRY OF ASTRONOMY. A Series of Familiar Essays. By R. A. PROCTOR. Crown 8vo. 6*s.*

NATURE STUDIES. By GRANT ALLEN, A. WILSON, T. FOSTER, E. CLODD, and R. A. PROCTOR. Crown 8vo. 6*s.*

LEISURE READINGS. By E. CLODD, A. WILSON, T. FOSTER, A. C. RUNYARD, and R. A. PROCTOR. Crown 8vo. 6*s.*

THE STARS IN THEIR SEASONS. An Easy Guide to a Knowledge of the Star Groups, in 12 Large Maps. By R. A. PROCTOR. Imperial 8vo. 5*s.*

STAR PRIMER. Showing the Starry Sky Week by Week, in 24 Hourly Maps. By R. A. PROCTOR. Crown 4to. 2*s.* 6*d.*

THE SEASONS PICTURED IN 48 SUN-VIEWS OF THE EARTH, and 24 Zodiacal Maps, &c. By R. A. PROCTOR. Demy 4to. 5*s.*

STRENGTH AND HAPPINESS. By R. A. PROCTOR. Crown 8vo. 5*s.*

ROUGH WAYS MADE SMOOTH. Familiar Essays on Scientific Subjects. By R. A. PROCTOR. Crown 8vo. 6*s.*

OUR PLACE AMONG INFINITIES. A Series of Essays contrasting our Little Abode in Space and Time with the Infinities Around us. By R. A. PROCTOR. Crown 8vo. 5*s.*

THE EXPANSE OF HEAVEN. Essays on the Wonders of the Firmament. By R. A. PROCTOR. Crown 8vo. 5*s.*

PLEASANT WAYS IN SCIENCE. By R. A. PROCTOR. Crown 8vo. 6*s.*

MYTHS AND MARVELS OF ASTRO-NOMY. By R. A. PROCTOR. Cr. 8vo. 6*s.*

Pryce. — *THE ANCIENT BRITISH CHURCH:* an Historical Essay. By JOHN PRYCE, M.A. Canon of Bangor. Crown 8vo. 6*s.*

Quain's Elements of Anatomy. The Ninth Edition. Re-edited by ALLEN THOMSON, M.D. LL.D. F.R.S.S. L. & E. EDWARD ALBERT SCHÄFER, F.R.S. and GEORGE DANCER THANE. With upwards of 1,000 Illustrations engraved on Wood, of which many are Coloured. 2 vols. 8vo. 18*s.* each.

Quain.—*A DICTIONARY OF MEDI-CINE.* By Various Writers. Edited by R. QUAIN, M.D. F.R.S. &c. With 138 Woodcuts. Medium 8vo. 31*s*. 6*d*. cloth, or 40*s*. half-russia; to be had also in 2 vols. 34*s*. cloth.

Reader.—*WORKS BY EMILY E. READER.*

THE GHOST OF BRANKINSHAW and other Tales. With 9 Full-page Illustrations. Fcp. 8vo. 2*s*. 6*d*. cloth extra, gilt edges.

VOICES FROM FLOWER-LAND, in Original Couplets. A Birthday-Book and Language of Flowers. 16mo. 1*s*. 6*d*. limp cloth; 2*s*. 6*d*. roan, gilt edges, or in vegetable vellum, gilt top.

FAIRY PRINCE FOLLOW-MY-LEAD; or, the *MAGIC BRACELET.* Illustrated by WM. READER. Crown 8vo. 2*s*. 6*d*. gilt edges; or 3*s*. 6*d*. vegetable vellum, gilt edges.

THE THREE GIANTS &c. Royal 16mo. 1*s*. cloth.

THE MODEL BOY &c. Royal 16mo. 1*s*. cloth.

BE YT HYS WHO FYNDS YT. Royal 16mo. 1*s*. cloth.

Reeve. — *COOKERY AND HOUSE-KEEPING.* By Mrs. HENRY REEVE. With 8 Coloured Plates and 37 Woodcuts. Crown 8vo. 7*s*. 6*d*.

Rich.—*A DICTIONARY OF ROMAN AND GREEK ANTIQUITIES.* With 2,000 Woodcuts. By A. RICH, B.A. Cr. 8vo. 7*s*. 6*d*.

Richardson.— *WORKS BY BENJAMIN WARD RICHARDSON, M.D.*

THE HEALTH OF NATIONS: a Review of the Works—Economical, Educational, Sanitary, and Administrative—of EDWIN CHADWICK, C.B. With a Biographical Dissertation by BENJAMIN WARD RICHARDSON, M.D. F.R.S. 2 vols. 8vo. 28*s*.

THE COMMONHEALTH: a Series of Essays on Health and Felicity for Every-Day Readers. Crown 8vo. 6*s*.

Richey.—*A SHORT HISTORY OF THE IRISH PEOPLE,* down to the Date of the Plantation of Ulster. By the late A. G. RICHEY, Q.C. LL.D. M.R.I.A. Edited, with Notes, by ROBERT ROMNEY KANE, LL.D. M.R.I.A. 8vo. 14*s*.

Riley.—*ATHOS;* or, the Mountain of the Monks. By ATHELSTAN RILEY, M.A. F.R.G.S. With Map and 29 Illustrations. 8vo. 21*s*.

Rivers. — *WORKS BY THOMAS RIVERS.*

THE ORCHARD-HOUSE. With 25 Woodcuts. Crown 8vo. 5*s*.

THE MINIATURE FRUIT GARDEN; or, the Culture of Pyramidal and Bush Fruit Trees, with Instructions for Root Pruning. With 32 Illustrations. Fcp. 8vo. 4*s*.

Robinson. — *THE NEW ARCADIA,* and other Poems. By A. MARY F. ROBINSON. Crown 8vo. 6*s*.

Roget.—*THESAURUS OF ENGLISH WORDS AND PHRASES,* Classified and Arranged so as to facilitate the Expression of Ideas and assist in Literary Composition. By PETER M. ROGET. Crown 8vo. 10*s*. 6*d*.

Ronalds. — *THE FLY-FISHER'S ENTOMOLOGY.* By ALFRED RONALDS. With 20 Coloured Plates. 8vo. 14*s*.

Saintsbury.—*MANCHESTER:* a Short History. By GEORGE SAINTSBURY. With 2 Maps. Crown 8vo. 3*s*. 6*d*.

Schäfer. — *THE ESSENTIALS OF HISTOLOGY, DESCRIPTIVE AND PRACTICAL.* For the use of Students. By E. A. SCHÄFER, F.R.S. With 281 Illustrations. 8vo. 6*s*. or Interleaved with Drawing Paper, 8*s*. 6*d*.

Schellen. — *SPECTRUM ANALYSIS IN ITS APPLICATION TO TERRESTRIAL SUBSTANCES,* and the Physical Constitution of the Heavenly Bodies. By Dr. H. SCHELLEN. Translated by JANE and CAROLINE LASSELL. Edited by Capt. W. DE W. ABNEY. With 14 Plates (including Angström's and Cornu's Maps) and 291 Woodcuts. 8vo. 31*s*. 6*d*.

Scott.—*WEATHER CHARTS AND STORM WARNINGS.* By ROBERT H. SCOTT, M.A. F.R.S. With numerous Illustrations. Crown 8vo. 6*s*.

Seebohm.—*WORKS BY FREDERIC SEEBOHM.*

THE OXFORD REFORMERS—JOHN COLET, ERASMUS, AND THOMAS MORE; a History of their Fellow-Work. 8vo. 14*s*.

THE ENGLISH VILLAGE COMMUNITY Examined in its Relations to the Manorial and Tribal Systems, &c. 13 Maps and Plates. 8vo. 16*s*.

THE ERA OF THE PROTESTANT REVOLUTION. With Map. Fcp. 8vo. 2*s*. 6*d*.

Sennett. — *THE MARINE STEAM ENGINE;* a Treatise for the use of Engineering Students and Officers of the Royal Navy. By RICHARD SENNETT, Engineer-in-Chief of the Royal Navy. With 244 Illustrations. 8vo. 21*s.*

Sewell. — *STORIES AND TALES.* By ELIZABETH M. SEWELL. Crown 8vo. 1*s.* each, boards ; 1*s.* 6*d.* each, cloth plain ; 2*s.* 6*d.* each, cloth extra, gilt edges :—

Amy Herbert.	Margaret Percival.
The Earl's Daughter.	Laneton Parsonage.
The Experience of Life.	Ursula.
A Glimpse of the World.	Gertrude.
Cleve Hall.	Ivors.
Katharine Ashton.	

Shakespeare. — *BOWDLER'S FAMILY SHAKESPEARE.* Genuine Edition, in 1 vol. medium 8vo. large type, with 36 Woodcuts, 14*s.* or in 6 vols. fcp. 8vo. 21*s.*

OUTLINES OF THE LIFE OF SHAKESPEARE. By J. O. HALLIWELL-PHILLIPPS, F.R.S. 2 vols. Royal 8vo. 10*s.* 6*d.*

Shilling Standard Novels.

BY THE EARL OF BEACONSFIELD.

Vivian Grey.	The Young Duke, &c.
Venetia.	Contarini Fleming,&c.
Tancred.	Henrietta Temple.
Sybil.	Lothair.
Coningsby.	Endymion.
Alroy, Ixion, &c.	

Price 1*s.* each, boards ; 1*s.* 6*d.* each, cloth.

BY G. J. WHYTE-MELVILLE.

The Gladiators.	Holmby House.
The Interpreter.	Kate Coventry.
Good for Nothing.	Digby Grand.
Queen's Maries.	General Bounce.

Price 1*s.* each, boards ; 1*s.* 6*d.* each, cloth.

BY ELIZABETH M. SEWELL.

Amy Herbert.	A Glimpse of theWorld.
Gertrude.	Ivors.
Earl's Daughter.	Katharine Ashton.
The Experience	Margaret Percival.
of Life.	Laneton Parsonage.
Cleve Hall.	Ursula.

Price 1*s.* each, boards ; 1*s.* 6*d.* each, cloth, plain ; 2*s.* 6*d.* each, cloth extra, gilt edges.

BY ANTHONY TROLLOPE.

The Warden.	Barchester Towers.

Price 1*s.* each, boards ; 1*s.* 6*d.* each, cloth.

BY ROBERT LOUIS STEVENSON.
  The Dynamiter.
  Strange Case of Dr. Jekyll and Mr. Hyde.
Price 1*s.* each, sewed ; 1*s.* 6*d.* each, cloth.

Shilling Standard Novels—*contd.*

BY BRET HARTE.
  In the Carquinez Woods. 1*s.* boards ; 1*s.* 6*d.* cloth.
  On the Frontier (Three Stories). 1*s.* sewed.
  By Shore and Sedge (Three Stories). 1*s.* sewed.

BY MRS. OLIPHANT.

In Trust.	Madam.

BY JAMES PAYN.
  Thicker than Water.
  The Luck of the Darrells.
Price 1*s.* each, boards ; 1*s.* 6*d.* each, cloth.

Short. — *SKETCH OF THE HISTORY OF THE CHURCH OF ENGLAND TO THE REVOLUTION OF 1688.* By T. V. SHORT, D.D. Crown 8vo. 7*s.* 6*d.*

Smith. — *LIBERTY AND LIBERALISM;* a Protest against the Growing Tendency toward Undue Interference by the State with Individual Liberty, Private Enterprise, and the Rights of Property. By BRUCE SMITH, of the Inner Temple, Barrister-at-Law. Crown 8vo. 6*s.*

Smith, H. F. — *THE HANDBOOK FOR MIDWIVES.* By HENRY FLY SMITH, M.B. Oxon. M.R.C.S. late Assistant-Surgeon at the Hospital for Sick Women, Soho Square. With 41 Woodcuts. Crown 8vo. 5*s.*

Smith, R. Bosworth. — *CARTHAGE AND THE CARTHAGINIANS.* By R. BOSWORTH SMITH, M.A. Maps, Plans, &c. Crown 8vo. 10*s.* 6*d.*

Smith, Rev. Sydney. — *THE WIT AND WISDOM OF THE REV. SYDNEY SMITH.* Crown 8vo. 1*s.* boards ; 1*s.* 6*d.* cloth.

Smith, T. — *A MANUAL OF OPERATIVE SURGERY ON THE DEAD BODY.* By THOMAS SMITH, Surgeon to St. Bartholomew's Hospital. A New Edition, re-edited by W. J. WALSHAM. With 46 Illustrations. 8vo. 12*s.*

Southey. — *THE POETICAL WORKS OF ROBERT SOUTHEY,* with the Author's last Corrections and Additions. Medium 8vo. with Portrait, 14*s.*

Stanley. — *A FAMILIAR HISTORY OF BIRDS.* By E. STANLEY, D.D. Revised and enlarged, with 160 Woodcuts. Crown 8vo. 6*s.*

Steel.—*WORKS BY J. H. STEEL, M.R.C.V.S.*

A TREATISE ON THE DISEASES OF THE DOG; being a Manual of Canine Pathology. Especially ¨adapted for the Use of Veterinary Practitioners and Students. 8vo. 10s. 6d.

A TREATISE ON THE DISEASES OF THE OX; being a Manual of Bovine Pathology specially adapted for the use of Veterinary Practitioners and Students. With 2 Plates and 117 Woodcuts. 8vo. 15s.

Stephen. — *ESSAYS IN ECCLESIASTICAL BIOGRAPHY.* By the Right Hon. Sir J. STEPHEN, LL.D. Crown 8vo. 7s. 6d.

Stevenson.—*WORKS BY ROBERT LOUIS STEVENSON.*

A CHILD'S GARDEN OF VERSES. Small fcp. 8vo. 5s.

THE DYNAMITER. Fcp. 8vo. 1s. swd. 1s. 6d. cloth.

STRANGE CASE OF DR. JEKYLL AND MR. HYDE. Fcp. 8vo. 1s. sewed; 1s. 6d. cloth.

'Stonehenge.' — *THE DOG IN HEALTH AND DISEASE.* By 'STONEHENGE.' With 84 Wood Engravings. Square crown 8vo. 7s. 6d.

THE GREYHOUND. By 'STONEHENGE.' With 25 Portraits of Greyhounds, &c. Square crown 8vo. 15s.

Stoney. — *THE THEORY OF THE STRESSES ON GIRDERS AND SIMILAR STRUCTURES.* With Practical Observations on the Strength and other Properties of Materials. By BINDON B. STONEY, LL.D. F.R.S. M.I.C.E. With 5 Plates, and 143 Illustrations in the Text. Royal 8vo. 36s.

Sully.—*WORKS BY JAMES SULLY.*

OUTLINES OF PSYCHOLOGY, with Special Reference to the Theory of Education. 8vo. 12s. 6d.

THE TEACHER'S HANDBOOK OF PSYCHOLOGY, on the Basis of 'Outlines of Psychology.' Crown 8vo. 6s. 6d.

Supernatural Religion; an Inquiry into the Reality of Divine Revelation. Complete Edition, thoroughly revised. 3 vols. 8vo. 36s.

Swinburne. — *PICTURE LOGIC;* an Attempt to Popularise the Science of Reasoning. By A. J. SWINBURNE, B.A. Post 8vo. 5s.

Taylor. — *STUDENT'S MANUAL OF THE HISTORY OF INDIA,* from the Earliest Period to the Present Time. By Colonel MEADOWS TAYLOR, C.S.I. Crown 8vo. 7s. 6d.

Taylor.—*AN AGRICULTURAL NOTE-BOOK:* to Assist Candidates in Preparing for the Science and Art and other Examinations in Agriculture. By W. C. TAYLOR. Crown 8vo. 2s. 6d.

Text-Books of Science: a Series of Elementary Works on Science, adapted for the use of Students in Public and Science Schools. Fcp. 8vo. fully illustrated with Woodcuts. *See p. 23.*

Thompson.—*WORKS BY D. GREENLEAF THOMPSON.*

THE PROBLEM OF EVIL: an Introduction to the Practical Sciences. 8vo. 10s. 6d.

A SYSTEM OF PSYCHOLOGY. 2 vols. 8vo. 36s.

Thomson's Conspectus.—Adapted to the British Pharmacopœia of 1885. Edited by NESTOR TIRARD, M.D. Lond. F.R.C.P. New Edition, with an Appendix containing notices of some of the more important non-official medicines and preparations. 18mo. 6s.

Thomson.—*AN OUTLINE OF THE NECESSARY LAWS OF THOUGHT;* a Treatise on Pure and Applied Logic. By W. THOMSON, D.D. Archbishop of York. Crown 8vo. 6s.

Three in Norway. By Two of THEM. With a Map and 59 Illustrations on Wood from Sketches by the Authors. Crown 8vo. 2s. boards; 2s. 6d. cloth.

Todd. — *ON PARLIAMENTARY GOVERNMENT IN ENGLAND:* its Origin, Development, and Practical Operation. By ALPHEUS TODD, LL.D. C.M.G. Librarian of Parliament for the Dominion of Canada. Second Edition, by his SON. In Two Volumes—VOL. I. 8vo. 24s.

Trevelyan.—*WORKS BY THE RIGHT HON. SIR G. O. TREVELYAN, BART.*

THE LIFE AND LETTERS OF LORD MACAULAY.

LIBRARY EDITION, 2 vols. 8vo. 36s.

CABINET EDITION, 2 vols. crown 8vo. 12s.

POPULAR EDITION, 1 vol. crown 8vo. 6s.

THE EARLY HISTORY OF CHARLES JAMES FOX. Library Edition, 8vo. 18s. Cabinet Edition, crown 8vo. 6s.

Trollope.—*NOVELS BY ANTHONY TROLLOPE.*

THE WARDEN. Crown 8vo. 1s. boards; 1s. 6d. cloth.

BARCHESTER TOWERS. Crown 8vo. 1s. boards; 1s. 6d. cloth.

Twiss.—*WORKS BY SIR TRAVERS TWISS.*

THE RIGHTS AND DUTIES OF NATIONS, considered as Independent Communities in Time of War. 8vo. 21s.

THE RIGHTS AND DUTIES OF NATIONS IN TIME OF PEACE. 8vo. 15s.

Tyndall.—*WORKS BY JOHN TYNDALL, F.R.S. &c.*

FRAGMENTS OF SCIENCE. 2 vols. crown 8vo. 16s.

HEAT A MODE OF MOTION. Crown 8vo. 12s.

SOUND. With 204 Woodcuts. Crown 8vo. 10s. 6d.

ESSAYS ON THE FLOATING-MATTER OF THE AIR in relation to Putrefaction and Infection. With 24 Woodcuts. Crown 8vo. 7s. 6d.

LECTURES ON LIGHT, delivered in America in 1872 and 1873. With 57 Diagrams. Crown 8vo. 5s.

LESSONS IN ELECTRICITY AT THE ROYAL INSTITUTION, 1875-76. With 58 Woodcuts. Crown 8vo. 2s. 6d.

NOTES OF A COURSE OF SEVEN LECTURES ON ELECTRICAL PHENOMENA AND THEORIES, delivered at the Royal Institution. Crown 8vo. 1s. sewed, 1s. 6d. cloth.

NOTES OF A COURSE OF NINE LECTURES ON LIGHT, delivered at the Royal Institution. Crown 8vo. 1s. sewed, 1s. 6d. cloth.

FARADAY AS A DISCOVERER. Fcp. 8vo. 3s. 6d.

Ville.—*ON ARTIFICIAL MANURES,* their Chemical Selection and Scientific Application to Agriculture. By GEORGES VILLE. Translated and edited by W. CROOKES, F.R.S. With 31 Plates. 8vo. 21s.

Virgil.—*PUBLI VERGILI MARONIS BUCOLICA, GEORGICA, ÆNEIS;* the Works of VIRGIL, Latin Text, with English Commentary and Index. By B. H. KENNEDY, D.D. Crown 8vo. 10s. 6d.

THE ÆNEID OF VIRGIL. Translated into English Verse. By J. CONINGTON, M.A. Crown 8vo. 9s.

THE POEMS OF VIRGIL. Translated into English Prose. By JOHN CONINGTON, M.A. Crown 8vo. 9s.

Vitzthum.—*ST. PETERSBURG AND LONDON IN THE YEARS* 1852-1864: Reminiscences of Count CHARLES FREDERICK VITZTHUM VON ECKSTOEDT, late Saxon Minister at the Court of St. James'. Edited, with a Preface, by HENRY REEVE, C.B. D.C.L. 2 vols. 8vo. 30s.

Walker. — *THE CORRECT CARD;* or, How to Play at Whist; a Whist Catechism. By Major A. CAMPBELL-WALKER, F.R.G.S. Fcp. 8vo. 2s. 6d.

Walpole.—*HISTORY OF ENGLAND FROM THE CONCLUSION OF THE GREAT WAR IN* 1815. By SPENCER WALPOLE. 5 vols. 8vo. Vols. I. and II. 1815-1832, 36s.; Vol. III. 1832-1841, 18s.; Vols. IV. and V. 1841-1858, 36s.

Waters. — *PARISH REGISTERS IN ENGLAND:* their History and Contents. With Suggestions for Securing their better Custody and Preservation. By ROBERT E. CHESTER WATERS, B.A. 8vo. 5s.

Watts.—*A DICTIONARY OF CHEMISTRY AND THE ALLIED BRANCHES OF OTHER SCIENCES.* Edited by HENRY WATTS, F.R.S. 9 vols. medium 8vo. £15. 2s. 6d.

Webb.—*CELESTIAL OBJECTS FOR COMMON TELESCOPES.* By the Rev. T. W. WEBB. Map, Plate, Woodcuts. Crown 8vo. 9s.

Wellington.—*LIFE OF THE DUKE OF WELLINGTON.* By the Rev. G. R. GLEIG, M.A. Crown 8vo. Portrait, 6s.

West.—*WORKS BY CHARLES WEST, M.D. &c.* Founder of, and formerly Physician to, the Hospital for Sick Children.

LECTURES ON THE DISEASES OF INFANCY AND CHILDHOOD. 8vo. 18s.

THE MOTHER'S MANUAL OF CHILDREN'S DISEASES. Crown 8vo. 2s. 6d.

Whately. — *ENGLISH SYNONYMS.* By E. JANE WHATELY. Edited by her Father, R. WHATELY, D.D. Fcp. 8vo. 3s.

Whately.—*WORKS BY R. WHATELY, D.D.*

ELEMENTS OF LOGIC. Crown 8vo. 4s. 6d.

ELEMENTS OF RHETORIC. Crown 8vo. 4s. 6d.

LESSONS ON REASONING. Fcp. 8vo. 1s. 6d.

BACON'S ESSAYS, with Annotations. 8vo. 10s. 6d.

White and Riddle.—*A LATIN-ENGLISH DICTIONARY.* By J. T. WHITE, D.D. Oxon. and J. J. E. RIDDLE, M.A. Oxon. Founded on the larger Dictionary of Freund. Royal 8vo. 21s.

White.—*A CONCISE LATIN-ENGLISH DICTIONARY*, for the Use of Advanced Scholars and University Students By the Rev. J. T. WHITE, D.D. Royal 8vo. 12s.

Wilcocks.—*THE SEA FISHERMAN.* Comprising the Chief Methods of Hook and Line Fishing in the British and other Seas, and Remarks on Nets, Boats, and Boating. By J. C. WILCOCKS. Profusely Illustrated. Crown 8vo. 6s.

Wilkinson.—*THE FRIENDLY SOCIETY MOVEMENT:* Its Origin, Rise, and Growth; its Social, Moral, and Educational Influences.—*THE AFFILIATED ORDERS.* —By the Rev. JOHN FROME WILKINSON, M.A. Crown 8vo. 2s. 6d.

Williams.—*PULMONARY CONSUMPTION;* its Etiology, Pathology, and Treatment. With an Analysis of 1,000 Cases to Exemplify its Duration and Modes of Arrest. By C. J. B. WILLIAMS, M.D. LL.D. F.R.S. F.R.C.P. and CHARLES THEODORE WILLIAMS, M.A. M.D.Oxon. F.R.C.P. With 4 Coloured Plates and 10 Woodcuts. 8vo. 16s.

Williams. — *MANUAL OF TELEGRAPHY.* By W. WILLIAMS, Superintendent of Indian Government Telegraphs. Illustrated by 93 Wood Engravings. 8vo. 10s. 6d.

Willich. — *POPULAR TABLES* for giving Information for ascertaining the value of Lifehold, Leasehold, and Church Property, the Public Funds, &c. By CHARLES M. WILLICH. Edited by H. BENCE JONES. Crown 8vo. 10s. 6d.

Wilson.—*A MANUAL OF HEALTH-SCIENCE.* Adapted for Use in Schools and Colleges, and suited to the Requirements of Students preparing for the Examinations in Hygiene of the Science and Art Department, &c. By ANDREW WILSON, F.R.S.E. F.L.S. &c. With 74 Illustrations. Crown 8vo. 2s. 6d.

Witt.—*WORKS BY PROF. WITT.* Translated from the German by FRANCES YOUNGHUSBAND.

THE TROJAN WAR. With a Preface by the Rev. W. G. RUTHERFORD, M.A. Head-Master of Westminster School. Crown 8vo. 2s.

MYTHS OF HELLAS; or, Greek Tales. Crown 8vo. 3s. 6d.

THE WANDERINGS OF ULYSSES. Crown 8vo. 3s. 6d.

Wood.—*WORKS BY REV. J. G. WOOD.*

HOMES WITHOUT HANDS; a Description of the Habitations of Animals, classed according to the Principle of Construction. With 140 Illustrations. 8vo. 10s. 6d.

INSECTS AT HOME; a Popular Account of British Insects, their Structure, Habits, and Transformations. With 700 Illustrations. 8vo. 10s. 6d.

INSECTS ABROAD; a Popular Account of Foreign Insects, their Structure, Habits, and Transformations. With 600 Illustrations. 8vo. 10s. 6d.

BIBLE ANIMALS; a Description of every Living Creature mentioned in the Scriptures. With 112 Illustrations. 8vo. 10s. 6d.

STRANGE DWELLINGS; a Description of the Habitations of Animals, abridged from 'Homes without Hands.' With 60 Illustrations. Crown 8vo. 5s. Popular Edition, 4to. 6d.

[*Continued on next page.*

Wood. — *WORKS BY REV. J. G. WOOD—continued.*

HORSE AND MAN: their Mutual Dependence and Duties. With 49 Illustrations. 8vo. 14*s.*

ILLUSTRATED STABLE MAXIMS. To be hung in Stables for the use of Grooms, Stablemen, and others who are in charge of Horses. On Sheet, 4*s.*

OUT OF DOORS; a Selection of Original Articles on Practical Natural History. With 11 Illustrations. Crown 8vo. 5*s.*

PETLAND REVISITED. With 33 Illustrations. Crown 8vo. 7*s. 6d.*

The following books are extracted from other works by the Rev. J. G. WOOD (*see* p. 21):

THE BRANCH BUILDERS. With 28 Illustrations. Crown 8vo. 2*s. 6d.* cloth extra, gilt edges.

WILD ANIMALS OF THE BIBLE. With 29 Illustrations. Crown 8vo. 3*s. 6d.* cloth extra, gilt edges.

DOMESTIC ANIMALS OF THE BIBLE. With 23 Illustrations. Crown 8vo. 3*s. 6d.* cloth extra, gilt edges.

BIRD-LIFE OF THE BIBLE. With 32 Illustrations. Crown 8vo. 3*s. 6d.* cloth extra, gilt edges.

WONDERFUL NESTS. With 30 Illustrations. Crown 8vo. 3*s. 6d.* cloth extra, gilt edges.

HOMES UNDER THE GROUND. With 28 Illustrations. Crown 8vo. 3*s. 6d.* cloth extra, gilt edges.

Wood-Martin. — *THE LAKE DWELLINGS OF IRELAND:* or Ancient Lacustrine Habitations of Erin, commonly called Crannogs. By W. G. WOOD-MARTIN, M.R.I.A. Lieut.-Colonel 8th Brigade North Irish Division, R.A. With 50 Plates. Royal 8vo. 25*s.*

Wright.—*HIP DISEASE IN CHILDHOOD,* with Special Reference to its Treatment by Excision. By G. A. WRIGHT, B.A. M.B.Oxon. F.R.C.S.Eng. With 48 Original Woodcuts. 8vo. 10*s. 6d.*

Wylie. — *HISTORY OF ENGLAND UNDER HENRY THE FOURTH.* By JAMES HAMILTON WYLIE, M.A. one of Her Majesty's Inspectors of Schools. (2 vols.) Vol. 1, crown 8vo. 10*s. 6d.*

Wylie. — *LABOUR, LEISURE, AND LUXURY;* a Contribution to Present Practical Political Economy. By ALEXANDER WYLIE, of Glasgow. Crown 8vo. 1*s.*

Youatt. — *WORKS BY WILLIAM YOUATT.*

THE HORSE. Revised and enlarged by W. WATSON, M.R.C.V.S. 8vo. Woodcuts, 7*s. 6d.*

THE DOG. Revised and enlarged. 8vo. Woodcuts. 6*s.*

Younghusband.—*THE STORY OF OUR LORD, TOLD IN SIMPLE LANGUAGE FOR CHILDREN.* By FRANCES YOUNGHUSBAND. With 25 Illustrations on Wood from Pictures by the Old Masters, and numerous Ornamental Borders, Initial Letters, &c. from Longmans' Illustrated New Testament. Crown 8vo. 2*s. 6d.* cloth plain; 3*s. 6d.* cloth extra, gilt edges.

Zeller. — *WORKS BY DR. E. ZELLER.*

HISTORY OF ECLECTICISM IN GREEK PHILOSOPHY. Translated by SARAH F. ALLEYNE. Crown 8vo. 10*s. 6d.*

THE STOICS, EPICUREANS, AND SCEPTICS. Translated by the Rev. O. J. REICHEL, M.A. Crown 8vo. 15*s.*

SOCRATES AND THE SOCRATIC SCHOOLS. Translated by the Rev. O. J. REICHEL, M.A. Crown 8vo. 10*s. 6d.*

PLATO AND THE OLDER ACADEMY. Translated by SARAH F. ALLEYNE and ALFRED GOODWIN, B.A. Crown 8vo. 18*s.*

THE PRE-SOCRATIC SCHOOLS; a History of Greek Philosophy from the Earliest Period to the time of Socrates. Translated by SARAH F. ALLEYNE. 2 vols. crown 8vo. 30*s.*

OUTLINES OF THE HISTORY OF GREEK PHILOSOPHY. Translated by SARAH F. ALLEYNE and EVELYN ABBOTT. Crown 8vo. 10*s. 6d.*

TEXT-BOOKS OF SCIENCE.

PHOTOGRAPHY. By Captain W. DE WIVE-LESLIE ABNEY, F.R.S. late Instructor in Chemistry and Photography at the School of Military Engineering, Chatham. With 105 Woodcuts. 3*s*.6*d*.

THE STRENGTH OF MATERIALS AND Structures : the Strength of Materials as depending on their quality and as ascertained by Testing Apparatus ; the Strength of Structures, as depending on their form and arrangement, and on the materials of which they are composed. By Sir J. ANDERSON, C.E. With 66 Woodcuts. 3*s*. 6*d*.

INTRODUCTION TO THE STUDY OF ORGANIC Chemistry ; the Chemistry of Carbon and its Compounds. By HENRY E. ARMSTRONG, Ph.D. F.R.S. With 8 Woodcuts. 3*s*. 6*d*.

ELEMENTS OF ASTRONOMY. By Sir R. S. BALL, LL.D. F.R.S. Andrews Professor of Astronomy in the Univ. of Dublin, Royal Astronomer of Ireland. With 136 Woodcuts. 6*s*.

RAILWAY APPLIANCES. A Description of Details of Railway Construction subsequent to the completion of Earthworks and Structures, including a short Notice of Railway Rolling Stock. By JOHN WOLFE BARRY, M.I.C.E. With 207 Woodcuts. 3*s*. 6*d*.

SYSTEMATIC MINERALOGY. By HILARY BAUERMAN, F.G.S. Associate of the Royal School of Mines. With 373 Woodcuts. 6*s*.

DESCRIPTIVE MINERALOGY. By HILARY BAUERMAN, F.G.S. Associate of the Royal School of Mines. With 236 Woodcuts. 6*s*.

METALS, THEIR PROPERTIES AND TREATment. By C. L. BLOXAM and A. K. HUNTINGTON, Professors in King's College, London. With 130 Woodcuts. 5*s*.

PRACTICAL PHYSICS. By R. T. GLAZEBROOK, M.A. F.R.S. and W. N. SHAW, M.A. With 80 Woodcuts. 6*s*.

PHYSICAL OPTICS. By R. T. GLAZEBROOK, M.A. F.R.S. Fellow and Lecturer of Trin. Coll. Demonstrator of Physics at the Cavendish Laboratory, Cambridge. With 183 Woodcuts of Apparatus, &c. 6*s*.

THE ART OF ELECTRO-METALLURGY, including all known Processes of Electro-Deposition. By G. GORE, LL.D. F.R.S. With 56 Woodcuts. 6*s*.

ALGEBRA AND TRIGONOMETRY. By WILLIAM NATHANIEL GRIFFIN, B.D. 3*s*. 6*d*.

NOTES ON THE ELEMENTS OF ALGEBRA and Trigonometry. With Solutions of the more difficult Questions. By W. N. GRIFFIN, B.D. 3*s*. 6*d*.

THE STEAM ENGINE. By GEORGE C. V. HOLMES, Whitworth Scholar ; Secretary of the Institution of Naval Architects. With 212 Woodcuts. 6*s*.

ELECTRICITY AND MAGNETISM. By FLEEMING JENKIN, F.R.SS. L. & E. late Professor of Engineering in the University of Edinburgh. With 177 Woodcuts. 3*s*. 6*d*.

THEORY OF HEAT. By J. CLERK MAXWELL, M.A. LL.D. Edin. F.R.SS. L. & E. With 41 Woodcuts. 3*s*. 6*d*.

TECHNICAL ARITHMETIC AND MENSURAtion. By CHARLES W. MERRIFIELD, F.R.S. 3*s*. 6*d*.

KEY TO MERRIFIELD'S TEXT-BOOK OF Technical Arithmetic and Mensuration. By the Rev. JOHN HUNTER, M.A. formerly Vice-Principal of the National Society's Training College, Battersea. 3*s*. 6*d*

INTRODUCTION TO THE STUDY OF INORganic Chemistry. By WILLIAM ALLEN MILLER, M.D. LL.D. F.R.S. With 72 Woodcuts. 3*s*. 6*d*.

TELEGRAPHY. By W. H. PREECE, F.R.S. M.I.C.E. and J. SIVEWRIGHT, M.A. C.M.G. With 160 Woodcuts. 5*s*.

THE STUDY OF ROCKS, an Elementary Text-Book of Petrology. By FRANK RUTLEY, F.G.S. of Her Majesty's Geological Survey. With 6 Plates and 88 Woodcuts. 4*s*. 6*d*.

WORKSHOP APPLIANCES, including Descriptions of some of the Gauging and Measuring Instruments—Hand Cutting Tools, Lathes, Drilling, Planing, and other Machine Tools used by Engineers. By C. P. B. SHELLEY, M.I.C.E. With 291 Woodcuts. 4*s*. 6*d*.

STRUCTURAL AND PHYSIOLOGICAL BOTANY. By Dr. OTTO WILHELM THOMÉ, Professor of Botany, School of Science and Art, Cologne, and A. W. BENNETT, M.A. B.Sc. F.L.S. With 600 Woodcuts. 6*s*.

QUANTITATIVE CHEMICAL ANALYSIS. By T. E. THORPE, Ph.D. F.R.S. Professor of Chemistry in the Andersonian University, Glasgow. With 88 Woodcuts. 4*s*. 6*d*.

QUALITATIVE ANALYSIS AND LABORATORY Practice. By T. E. THORPE, Ph.D. F.R.S.E Professor of Chemistry in the Andersonian University, Glasgow ; and M. M. PATTISON MUIR. With Plate of Spectra and 57 Woodcuts. 3*s*. 6*d*.

INTRODUCTION TO THE STUDY OF CHEMical Philosophy ; the Principles of Theoretical and Systematical Chemistry. By WILLIAM A. TILDEN, D.Sc. London, F.R.S. With 5 Woodcuts. 3*s*. 6*d*. With Answers to Problems, 4*s*. 6*d*.

ELEMENTS OF MACHINE DESIGN ; an Introduction to the Principles which determine the Arrangement and Proportion of the Parts of Machines, and a Collection of Rules for Machine Designs. By W. CAWTHORNE UNWIN, B.Sc. M.I.C.E. With 325 Woodcuts. 6*s*.

PLANE AND SOLID GEOMETRY. By H. W. WATSON, M.A. formerly Fellow of Trinity College, Cambridge. 3*s*. 6*d*.

EPOCHS OF ANCIENT HISTORY.

Edited by the Rev. Sir G. W. Cox, Bart. M.A. and by C. SANKEY, M.A. 10 Volumes, fcp. 8vo. with numerous Maps, Plans, and Tables, price 2s. 6d. each volume.

THE GRACCHI, MARIUS, AND SULLA. By A. H. BEESLY, M.A.

THE EARLY ROMAN EMPIRE. From the Assassination of Julius Cæsar to the Assassination of Domitian. By the Rev. W. WOLFE CAPES, M.A.

THE ROMAN EMPIRE OF THE SECOND CENtury, or the Age of the Antonines. By the Rev. W. WOLFE CAPES, M.A.

THE ATHENIAN EMPIRE. From the Flight of Xerxes to the Fall of Athens. By the Rev. Sir G. W. Cox, Bart. M.A.

THE GREEKS AND THE PERSIANS. By the Rev. Sir G. W. Cox, Bart. M.A.

THE RISE OF THE MACEDONIAN EMPIRE. By ARTHUR M. CURTEIS, M.A.

ROME TO ITS CAPTURE BY THE GAULS. By WILHELM IHNE.

THE ROMAN TRIUMVIRATES. By the Very Rev. CHARLES MERIVALE, D.D.

THE SPARTAN AND THEBAN SUPREMACIES. By CHARLES SANKEY, M.A.

ROME AND CARTHAGE, THE PUNIC WARS. By R. BOSWORTH SMITH, M.A.

EPOCHS OF MODERN HISTORY.

Edited by C. COLBECK, M.A. 18 vols. fcp. 8vo. with Maps, price 2s. 6d. each volume.

THE NORMANS IN EUROPE. By Rev. A. H. JOHNSON, M.A.

THE CRUSADES. By the Rev. Sir G. W. Cox, Bart. M.A.

THE BEGINNING OF THE MIDDLE AGES. By R. W. CHURCH, D.D. Dean of St. Paul's.

THE EARLY PLANTAGENETS. By W. STUBBS, D.D. Bishop of Chester.

EDWARD THE THIRD. By the Rev. W. WARBURTON, M.A.

THE HOUSES OF LANCASTER AND YORK. By JAMES GAIRDNER.

THE EARLY TUDORS. By the Rev. C. E. MOBERLY, M.A.

THE ERA OF THE PROTESTANT REVOLUtion. By F. SEEBOHM.

THE FIRST TWO STUARTS AND THE PURItan Revolution, 1603–1660. By SAMUEL RAWSON GARDINER.

THE AGE OF ELIZABETH. By the Rev. M. CREIGHTON, M.A. LL.D.

THE FALL OF THE STUARTS; AND WESTERN Europe from 1678 to 1697. By the Rev. EDWARD HALE, M.A.

THE AGE OF ANNE. By E. E. MORRIS, M.A.

THE THIRTY YEARS' WAR, 1618–1648. By SAMUEL RAWSON GARDINER.

THE EARLY HANOVERIANS. By E. E. MORRIS, M.A.

FREDERICK THE GREAT AND THE SEVEN Years' War. By F. W. LONGMAN.

THE WAR OF AMERICAN INDEPENDENCE, 1775–1783. By J. M. LUDLOW.

THE FRENCH REVOLUTION, 1789–1795. By Mrs. S. R. GARDINER.

THE EPOCH-OF REFORM, 1830–1850. By JUSTIN M'CARTHY, M.P.

EPOCHS OF ENGLISH HISTORY.

Edited by the Rev. MANDELL CREIGHTON, M.A.

EARLY ENGLAND TO THE NORMAN CONquest. By F. YORK POWELL, M.A. 1s.

ENGLAND A CONTINENTAL POWER, 1066–1216. By Mrs. MANDELL CREIGHTON. 9d.

RISE OF THE PEOPLE AND THE GROWTH OF Parliament, 1215–1485. By JAMES ROWLEY, M.A. 9d.

TUDORS AND THE REFORMATION, 1485–1603. By the Rev. MANDELL CREIGHTON. 9d.

STRUGGLE AGAINST ABSOLUTE MONARCHY, 1603–1688. By Mrs. S. R. GARDINER. 9d.

SETTLEMENT OF THE CONSTITUTION, from 1689 to 1784. By JAMES ROWLEY, M.A. 9d.

ENGLAND DURING THE AMERICAN AND European Wars, from 1765 to 1820. By the Rev. O. W. TANCOCK, M.A. 9d.

MODERN ENGLAND FROM 1820 TO 1874. By OSCAR BROWNING, M.A. 9d.

*** Complete in One Volume, with 27 Tables and Pedigrees, and 23 Maps. Fcp. 8vo. 5s.

THE SHILLING HISTORY OF ENGLAND; being an Introductory Volume to the Series of 'Epochs of English History.' By the Rev. MANDELL CREIGHTON, M.A. Fcp. 8vo. 1s.

EPOCHS OF CHURCH HISTORY.

Edited by the Rev. MANDELL CREIGHTON, M.A. Fcp. 8vo. price 2s. 6d. each volume.

THE ENGLISH CHURCH IN OTHER LANDS; or, the Spiritual Expansion of England. By Rev. W. H. TUCKER, M.A.

THE HISTORY OF THE REFORMATION IN England. By GEORGE G. PERRY, M.A.

THE EVANGELICAL REVIVAL IN THE Eighteenth Century. By the Rev. JOHN HENRY OVERTON, M.A.

THE CHURCH OF THE EARLY FATHERS. External History. By ALFRED PLUMMER, M.A. D.D.

THE HISTORY OF THE UNIVERSITY OF Oxford. By the Hon. G. C. BRODRICK, D.C.L.

THE CHURCH AND THE ROMAN EMPIRE. By the Rev. ARTHUR CARR, M.A.

THE CHURCH AND THE PURITANS, 1570–1660. By HENRY OFFLEY WAKEMAN, M.A.

*** *Other Volumes in preparation.*

Spottiswoode & Co. Printers, New-street Square, London.

www.ingramcontent.com/pod-product-compliance
Lightning Source LLC
Chambersburg PA
CBHW021512210326
41599CB00012B/1230